AGRICULTURE ISSUES AND POLICIES

FARMS IN THE UNITED STATES

SIZE, STRUCTURE AND FORCES OF CHANGE

AGRICULTURE ISSUES AND POLICIES

Additional books in this series can be found on Nova's website under the Series tab.

Additional e-books in this series can be found on Nova's website under the e-book tab.

AGRICULTURE ISSUES AND POLICIES

FARMS IN THE UNITED STATES

SIZE, STRUCTURE AND FORCES OF CHANGE

JANINE GROSSO
EDITOR

nova publishers
New York

Library of Congress Cataloging-in-Publication Data

ISBN: 978-1-63483-667-8

Published by Nova Science Publishers, Inc. † New York

CONTENTS

PREFACE

Large farms now dominate crop production in the United States. Although most cropland was operated by farms with less than 600 crop acres in the early 1980s, today most cropland is on farms with at least 1,100 acres, and many farms are 5 and 10 times that size. This ongoing shift in farm structure raises a host of questions. How extensive is this structural change? What forces have accompanied and contributed to the shift to larger farms? What implications do these structural shifts have for family farms? The shift of acreage to larger farms is part of a complex set of structural changes in crop agriculture. The number of mid-size crop farms has declined, while farm numbers at the extremes (large and small) are growing. Because of these changes, average farm size has changed little in the last three decades, even while cropland and crop production have shifted to much larger farms. This book documents the complex nature of modern farm structure and introduces statistics aimed at better tracking consolidation of land and production. Furthermore, this book looks at small acreage farms having gross sales of $10,000 or more in a given year to better understand the product choices and strategies used by small acreage farms that appear to be operating profitably.

In: Farms in the United States
Editor: Janine Grosso
ISBN: 978-1-63483-667-8

Chapter 1

FARM SIZE AND THE ORGANIZATION OF U.S. CROP FARMING*

James M. MacDonald, Penni Korb and Robert A. Hoppe

ABSTRACT

Cropland has been shifting to larger farms. The shifts have been large, centered on a doubling of farm size over 20-25 years, and they have been ubiquitous across States and commodities. But the shifts have also been complex, with land and production shifting primarily from mid-size commercial farming operations to larger farms, while the count of very small farms increases. Larger crop farms still realize better financial returns, on average, and they are able to make more intensive use of their labor and capital resources, indicating that the trends are likely to continue. The report relies on comprehensive farm-level data to detail changes in farm size and other attributes of farm structure, and to evaluate the key driving forces, inclu ding technologies, farm organization and business relationships, land attributes, and government policies.

* This is an edited, reformatted and augmented version of Economic Research Report Number 152, issued by the U.S. Department of Agriculture, Economic Research Service, August 2013.

Errata

This report, originally released in August 2013, was reissued with the following correction:

On September 9, 2013, a citation was added to the report's list of references (U.S. Department of Agriculture, National Agricultural Statistics Service (NASS). 2009. Farms, Land in Farms, and Livestock Operations 2008 Summary. February), completing a reference to "USDA (2009)" in footnote 3 in the box on page 6.

What Is the Issue?

Large farms now dominate crop production in the United States. Although most cropland was operated by farms with less than 600 crop acres in the early 1980s, today most cropland is on farms with at least 1,100 acres, and many farms are 5 and 10 times that size. This ongoing shift in farm structure raises a host of questions. How extensive is this structural change? What forces have accompanied and contributed to the shift to larger farms? What implications do these structural shifts have for family farms?

The shift of acreage to larger farms is part of a complex set of structural changes in crop agriculture. The number of mid-size crop farms has declined, while farm numbers at the extremes (large and small) are growing. Because of these changes, average farm size has changed little in the last three decades, even while cropland and crop production have shifted to much larger farms. The report documents the complex nature of modern farm structure and introduces statistics aimed at better tracking consolidation of land and production.

What Did the Study Find?

The report introduces a measure of midpoint acreage in which half of all *cropland acres* are on farms with more cropland than the midpoint, and half are on farms with less. Midpoint acreage is revealed to be a more informative measure of cropland consolidation than either a simple median (in which half of all *farms* are either larger or smaller) or the simple mean (which is average cropland per farm). Using this measure, ERS researchers found that:

- The midpoint acreage for U.S. cropland nearly doubled between 1982 and 2007, from 589 acres to 1,105.
- Midpoint acreages increased in 45 of 50 States and more than doubled in 16. The largest increases occurred in a contiguous group of 12 Corn Belt and Northern Plains States.
- Midpoint acreages more than doubled in each of 5 major field crops (corn, cotton, rice, soybeans, and wheat) and increased in 35 of 39 fruit and vegetable crops, where the average increase was 107 percent.
- The shifts were persistent, with a general pattern of increase between each Census of Agriculture conducted between 1982 and 2007. However, less comprehensive evidence from annual surveys suggests that the pace of consolidation slowed between 2007 and 2011, the last year for which data are available. Data from the 2012 Census will provide more definitive evidence of recent trends.
- Larger crop farms continue to realize better financial performance: average rates of return on equity increased with farm size in five major commodity categories analyzed in this report (corn, soybeans, wheat, fruits, and vegetables). In turn, larger farms utilize labor and capital more intensively, which provide them with the primary source of their financial advantage.

The long-term shifts in farm size have been accompanied by greater specialization—beginning with a separation of livestock farming from crop farming in the latter half of the 20th century. As crop and livestock production separated, full-time crop farmers could devote more time to crop production and manage more cropland. At the same time, the number of production and marketing contracts to govern the sale of products has increased. Contracts covered 32 percent of crop production in 2011, compared with 23 percent in the mid-1990s. Larger operations are more likely to use contracts, which can reduce the price and marketing risks faced by farmers.

Technology also plays an important role in driving increases in farm size, by allowing a single farmer to operate and manage more acres. Labor-saving innovations—from bigger and faster capital equipment to information technology, chemical herbicides, seed genetics, and changing tillage techniques—have substantially reduced the total amount of labor used in agriculture and facilitated the shift to larger crop farms.

Federal policies may affect farm structure through multiple channels, such as taxes, lending programs, environmental or food safety regulation, research and development funding, and commodity programs. Some effects are straightforward and fairly direct, while others are subtle and indirect. The impact of broad commodity policy is particularly complex. Some have argued that commodity and crop insurance programs, by reducing the financial risks faced by farmers, encouraged the adoption of organizational forms and capital equipment that spurred increases in farm size. It is, however, difficult to separate the impact of policy from technology itself and from other factors that affected farm risks.

While the above conditions may have facilitated the shifts toward larger farms, family farms continue to dominate crop agriculture. In 2011, 96 percent of U.S. crop farms were family farms, and they accounted for 87 percent of the value of crop production.

How Was the Study Conducted?

This study drew upon data from two main sources. The Census of Agriculture, conducted by the USDA's National Agricultural Statistics Service (NASS), provides comprehensive, historical, and publicly available data on consolidation and specialization trends. The study also relied on confidential farm-level data from the census accessed in a secure environment to ensure confidentiality, to generate measures of consolidation and farm size for the United States, and major commodities for 1982-2007.

The second primary source of data is the annual Agricultural Resource Management Survey (ARMS), jointly administered by NASS and ERS. The ARMS covers U.S. farming operations and their operators in the 48 contiguous States. The survey was used to supplement historic census data on consolidation with more recent annual developments, to provide data on financial performance among crop farms, to assess the role of family farms, and to provide evidence on the use of labor, capital, and various production practices among crop farms.

The study also used several additional datasets and publications from NASS, ERS, and other Government and private sources.

Introduction

Large farms now dominate crop production in the United States. While most cropland was on farms with less than 600 acres in the early 1980s, today most cropland is on farms with at least 1,100 acres, and many crop farms are 5 and 10 times that size. This ongoing shift in control of farm resources and production raises a host of questions addressed in this report. How extensive is this structural change? Why is cropland moving to larger farms? Is the shift associated with other changes in farm structure, and will it continue? Do these structural shifts portend an end to family farms, and the emergence of corporate farms with diffused ownership and hired professional managers?

Economic Research Service (ERS) researchers measure and describe the nature of the changes that have occurred and assess the driving forces behind those changes, with an emphasis on technology, land attributes, farm organization, and public policy.[1] The shift of acreage and production to larger farms is ongoing and ubiquitous, occurring in most States and for most crops, including major field crops as well as fruits and vegetables. Because large farms continue to realize better financial returns, on average, than small and mid-size farms, it is reasonable to expect continued shifts of production and land to larger farms.

The complexity of U.S. farm structure makes the shift of acreage to larger farms hard to discern with common metrics. In recent years, the number of mid-size crop farms has declined sharply, while farm numbers at the extremes (large and small) have grown. As a result the size of the average farm (in acres) has changed little, but large farms have gotten considerably larger as they farm more of the country's cropland. The report documents the complex nature of modern farm structure and introduces statistics aimed at better tracking shifts of land and production.

Measuring Farm Size

In this report, farm size is defined according to acres of cropland operated by the farm, which is a transparent, easily understood measure for which statistics have been kept for many years. It is particularly well suited to the primary purpose of the report, assessing the consolidation of cropland into larger operations over time.

Farms may own cropland, but they also grow crops on cropland that they rent. The report defines farm size according to the cropland operated by the farm—that is, the cropland it owns, plus any that it rents, minus any rented to others.

All cropland is not the same. Across regions, important differences exist in soil quality, topography, climatic conditions, and proximity to markets and urban areas. These differences affect the types of crops grown, the sales derived from a given amount of acreage, and the size of crop farms. To account for some of those differences in cropland attributes, ERS researchers measured and analyzed changes in farm size for States and for specific crops, in addition to national data.

About one-quarter of cropland is not harvested, and most of that is not planted to any crop.[2] Analyses of specific commodities rely on harvested acreage of the commodity because the data sources report acres harvested, rather than acres planted, on a consistent basis, while analyses of farm size include both harvested and non-harvested cropland. (See box: "Data, Current and Historic.")

Data, Current and Historic

ERS researchers supplemented Agricultural Resource Management Survey (ARMS) and Census of Agriculture data with information from several other National Agricultural Statistics Service (NASS) surveys, from ERS estimates, and from several other public and private sources. These sources are described in the report as they are introduced.

Census of Agriculture data. The report draws on publicly available data from the Census of Agriculture, administered by NASS. The census elicits information from all U.S. farms and is valuable for following long-term trends, since it dates back to 1840. It provides deep and comprehensive coverage of acreage and production for all U.S. States and for a wide range of commodities. The census was conducted at irregular 4-, 5-, and 10-year intervals until 1982 and at 5-year intervals since then. Information on the census can be found at www.agcensus.usda.gov/index.php.

However, besides publicly available census data, ERS researchers also used confidential farm-level census records for 1982-2007, accessed under an agreement with NASS that is designed to protect data security and confidentiality. The data were used to develop improved measures of consolidation for the Nation, States, and commodities.

ARMS data. The research also relies heavily on the Agricultural Resource Management Survey (ARMS), an annual survey of U.S. farms that is jointly administered by NASS and ERS. The ARMS is based on a representative sample of farms and their operators in the 48 contiguous States.

The survey has been conducted annually since 1996, and the most recent data available for this report covered farm performance in 2011.

With a total national sample of just over 20,000 farms, ARMS cannot provide the comprehensive measures at the State and commodity level that the census provides, and it cannot match the long temporal span of the census. However, the survey gives more detailed data on farm finances, operators, resources, and practices than are available elsewhere. As an annual survey, it also gives greater and more recent temporal detail than the census.

ARMS comprises three phases, carried out at different times of the year. Phase I is a screening module, not used for research. Phase II, conducted in the fall of the reference year, targets up to two field crops. It focuses on field-level information, with questions on chemical use, resources and input use, production practices, and production outcomes. ERS researchers used Phase II data for a few selected topics, including tillage practices, information technology, and seed use. Most of the analyses are drawn from Phase III, which is conducted in the winter following the reference year and is aimed at all farms. Phase III focuses on production, input use, farm and operator attributes, and finances for the whole farm.

During most of the years covered in this report, Phase III includes five questionnaire versions. Version 5, shorter than the others, contains a core set of questions. Version 1 adds a detailed set of research questions on production and management practices and resource use to the core. Versions 2, 3, and 4—known as commodity versions—add commodity-specific questions to the core and are aimed at representative samples of commercial producers of the Phase II crops and one livestock commodity.

Most ARMS-based statistics in this report are drawn from the core questions that appear on all versions of the survey, while some are based on questions that appear only in Version 1.[1] As necessary, observations from several years are combined to generate larger samples (ARMS is not a panel, so different farms appear in different years). In the text discussion and table notes, distinctions are drawn between:

- Data drawn from Phase II and from Phase III;
- Data drawn from all Phase III versions and from version 1 only;
- Farms from a single year and farms pooled across 2008-2011.

Finally, NASS uses a stratified sampling strategy to improve the reliability of estimates in ARMS. That means that some farms have a higher probability of sample selection—larger operations are more likely to be selected than smaller ones, and selection probabilities also vary across geographic areas and commodity types. Population estimates are generated by weighting sample observations to reflect their varying selection probabilities. Further information on ARMS, including questionnaires, can be found at www.ers.usda.gov/data-products/arms-farm-financial-and-crop-production-practices.aspx.

[1]The survey is designed so that Version 1 records can be expanded to represent all farms; a separate set of weights allows the all-version sample to be expanded to represent all farms. Each commodity version carries additional weights to allow for expansion to represent the population of commercial producers of each commodity.

CONSOLIDATION OF CROPLAND INTO LARGER FARMS

The average size of a U.S. crop farm has changed little during the past three decades. However, this seeming stability masks important structural changes in the complex U.S. farm sector. There are growing numbers of very small and very large farms and declining numbers of mid-sized farms. Cropland acreage has moved toward much larger farms.

Start with the complex size pattern of U.S. farms. In figure 1, based on 2011 ARMS data, farms and cropland acreage are sorted into eight cropland size classes frequently reported in Census of Agriculture publications. In that year, 391.6 million acres of cropland were divided among 1.675 million U.S. farms with cropland, for an average (mean) farm size of 234 acres. However, relatively few farms are near the average. Eighty percent of farms with cropland were smaller than the mean size, and 70 percent were less than half the mean size.

The median farm size (at which half of farms were larger and half were smaller) was just 45 acres.

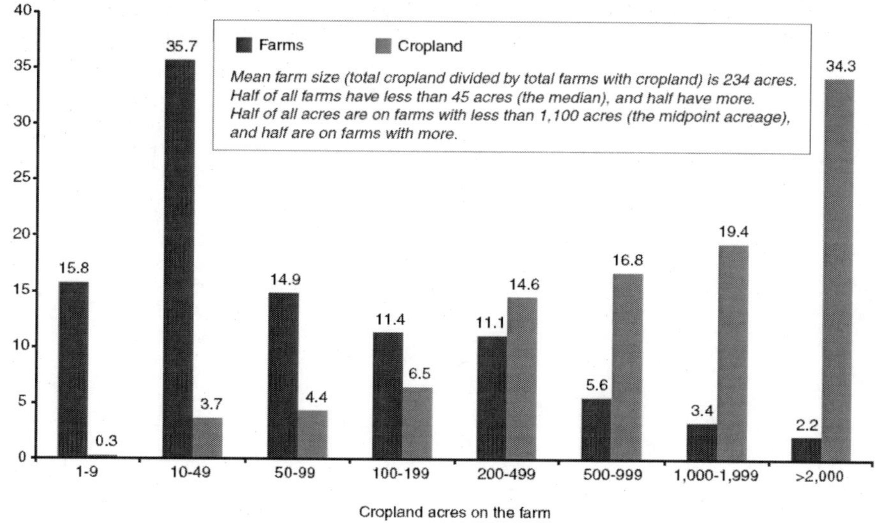

Note: Farm size is defined according to the cropland the farm operates—that is, the cropland it owns, plus any that it rents, minus any rented to others.
Source: USDA Agricultural Resource Management Survey, 2011.

Figure 1. The size distribution of crop farms, 2011.

Similarly, little cropland is on farms near the average. Eighty-three percent of cropland was on farms that were larger than the mean size, and 71 percent was on farms that were more than twice the mean. The midpoint acreage—where half of cropland is on larger farms and half on smaller— was 1,100 acres.

Figure 2 reports the same data for 2001; together the two figures summarize a decade of structural change. The mean farm size was little different (235 acres), but the median farm size in 2001 (63 acres) was substantially larger than that in 2011. There were nearly 100,000 more farms with 1-49 acres of cropland in 2011 than in 2001, as the count of small farms in USDA statistics increased sharply. (See box: "The Increasing Number of Small Crop Farms.") Cropland moved in the other direction: the largest farms (at least 2,000 acres of cropland) accounted for 34.3 percent of cropland in 2011, up from 24.1 percent in 2001, and the number of farms with at least 2,000 acres of cropland increased during the decade. The midpoint acreage reflects the shift in cropland: it was 900 acres in 2001 (fig. 2), compared to 1,100 in 2011 (fig. 1).

Because of the complexity of changes in crop farm structure, simple measures of mean farm size are not very informative. Simple means and medians focus on the average farm, and the land operated by the average farmer. This report is focused on the use of cropland and must focus on the average acre of cropland, not the average farmer or average farm. The midpoint acreage effectively tracks cropland consolidation and will be used in this report.

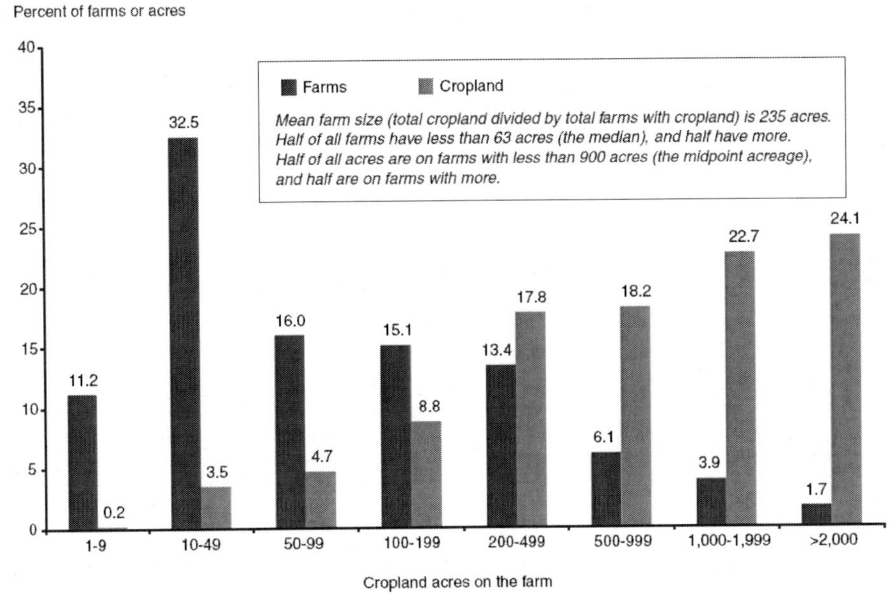

Note: Farm size is defined according to the cropland the farm operates—that is, the cropland it owns, plus any that it rents, minus any rented to others.
Source: USDA Agricultural Resource Management Survey, 2001.

Figure 2. The size distribution of U.S. crop farms, 2001.

The Increasing Number of Small Crop Farms

According to the Census of Agriculture, 872,000 farms harvested 1-49 acres of cropland in 1974, a decline of 70 percent from the 2.9 million farms in that size class in 1950. If the 1950- 1974 trend had continued, there would have been less than 150,000 farms in that size class by 2011.

However, the trend did not continue; the number of small farms, whether measured by sales or by acreage, declined much less rapidly after 1974 and began to increase after 2000.

In part, the survival and expansion of small crop farms may reflect farming opportunities as well as popular lifestyle choices by people who prefer to combine a rural lifestyle with modest crop or livestock production. A farmer can operate a small field crop operation on a part-time basis while also maintaining an off-farm job, or while drawing additional income from pensions and savings.[1]

However, how farms are counted also matters when considering trends in farm numbers. USDA, under the direction of Congress, has since 1974 defined a farm as a place that produces, or normally could produce, at least $1,000 worth of agricultural commodities in a year.[2] Because the definition is not adjusted for inflation, it will define more very small places as farms as farm commodity prices rise, and the Producer Price Index for Farm Products (Bureau of Labor Statistics) rose by 172 percent between 1974, when the definition was set, and 2011.

USDA/NASS has also made concerted efforts in recent years to better identify and track very small farms.[3] While these efforts improve survey methodology and provide more comprehensive coverage, their implementation indicates that some part of the observed increase in small farm numbers reflects better counting, not more farms. Because most of these farms are so small, with very limited production, their inclusion adds very little cropland or production to census aggregates.

[1] By pooling ARMS Phase III records from 2008-2011, ERS analysts constructed a large sample of smaller, but still commercial, crop farms with 50-99 acres of grain or oilseed crops, and no livestock production. On average, operators of those farms report working 20 hours per week on the farm, and smaller operations could require less labor time.

[2] A place with less than $1,000 in sales in a year is classified as a farm if it has cropland or animal assets that could generate $1,000 in sales (the "normally could produce" part of the farm definition). There were 254,000 such farms in 1982, 232,000 in 1992, and 689,000 in 2007, a near-tripling in 15 years. For more on defining farms, see O'Donoghue, et al. (2009).

[3] For more on expanded efforts to track small farms, see the "Special Note Regarding the 2007 Estimates" on p. 31 in U.S. Department of Agriculture (2009).

The midpoint acreage is a median, but it differs from the median farm size that is also reported in figures 1 and 2. Technically, the midpoint acreage is the median of the distribution of *acreage* by farm size, as opposed to the more commonly reported median of the distribution of *farms* by farm size. (See box: "Midpoint measures in industry analyses.") The median farm size and the midpoint acreage differ widely: the 2011 median farm size was 45 acres, while the midpoint acreage was 1,100 acres, because they measure different concepts and because U.S. farms cover such a wide range of sizes.

Midpoint Measures in Industry Analyses

Midpoints have been widely used in industry analyses, often based on employment, where half of industry employees are at larger plants or firms, and half are at smaller (Florence, 1933; Scherer and Ross, 1990; Davis and Haltiwanger, 1991). As distinct from simple medians, they have been called Florence medians, referring to the first economics author to use them in 1933, and weighted medians, since they can be calculated by weighting each observation by its size (acreage, or employment in the most common applications). The measure is especially useful for size distributions that are highly skewed, with many very small operations while employment, acreage, or production is concentrated in a small number of large firms. Most U.S industries, including agriculture, are characterized by highly skewed size distributions. Acreage-based midpoints have been applied to the measurement of farm size by Lund and Price (1998), Key and Roberts (2007a), and O'Donoghue et al. (2011). Lund and Price (1998) coined the midpoint usage.

ARMS data were used in figures 1 and 2 to summarize modern-day farm structure. ERS analysts used Census of Agriculture data to compare longer run trends in midpoint acreages and simple means during 1982-2007 (fig. 3).[3] The differences are striking. While mean farm size changed little between 1982 and 2007—from 221 acres to 241, or a 9-percent increase—the midpoint for cropland grew by 88 percent, from 589 acres to 1,105. The simple mean peaked in 1997, and declined thereafter, but the midpoint acreage rose steadily from 1982 through 2007. The differences reflect the shift of cropland to larger farms, even as increases in the number of very small farms kept the mean farm size from increasing substantially. While the average farm did not get much larger, acreage moved to much larger farms.[4]

The trend in the midpoint for harvested cropland does not exactly match that for cropland—the cropland measure rises steadily from 1982 to 2007, while the harvested cropland measure only rises after 1987—but the basic message is similar (fig. 3). The midpoint acreage for harvested cropland rose by 114 percent, from 500 acres in 1982 to 1,071 acres in 2007.

Shifts of cropland to larger farms mirror similar shifts of agricultural production from smaller to larger sales classes during 1982-2007. Hoppe et al. (2010) use Census of Agriculture data to show that farms with at least $1 million in sales (measured in 2007 dollars, and thus adjusted for changes in prices) accounted for 24 percent of the value of agricultural production in 1982 and 59 percent in 2007. Meanwhile the share held by small commercial farms, with $10,000 to $250,000 in sales, fell by two-thirds.

Cropland Consolidation in the States

The pace of cropland consolidation varied widely across States, although most saw substantial increases. Figure 4 reports State-by-State changes in the midpoint acreages for harvested cropland between 1982 and 2007. Harvested cropland shifted to smaller farms in 5 States, but shifted to larger farms in the other 45, and midpoint acreages rose by more than 100 percent in 16 States. Declines occurred in Hawaii, where plantation agriculture that focused on pineapple and banana production was shifting to more diversified crop agriculture.

Declines also occurred in Alabama and Southern New England. Increases of at least 100 percent occurred across a contiguous set of 12 Corn Belt and Northern Great Plains States, as well as Idaho, Maryland, Mississippi, and North Carolina.

A simple statistical analysis helps to summarize the patterns. Specifically, four attributes of a State's agricultural environment account for a substantial share of the variation in growth rates across States. (See box: "Accounting for Cross-State Differences in Consolidation.") Midpoint acreages increased more in areas where most land was cropland, with little land in forest, pasture, commercial, or residential uses. In such States, crop fields are more likely to be large and contiguous to one another, rather than scattered among other land uses. Consolidation also increased faster where population was less dense; denser populations make cropland consolidation more costly because of greater competition for land.

Finally, midpoint acreages increased more in those States with high shares of harvested cropland relative to cropland, and where hay accounted for a smaller share of harvested cropland. Hay was more likely to be produced in regions with noncontiguous cropland and smaller fields, and it was also concentrated on hillier land that is often less suited to other field crops.

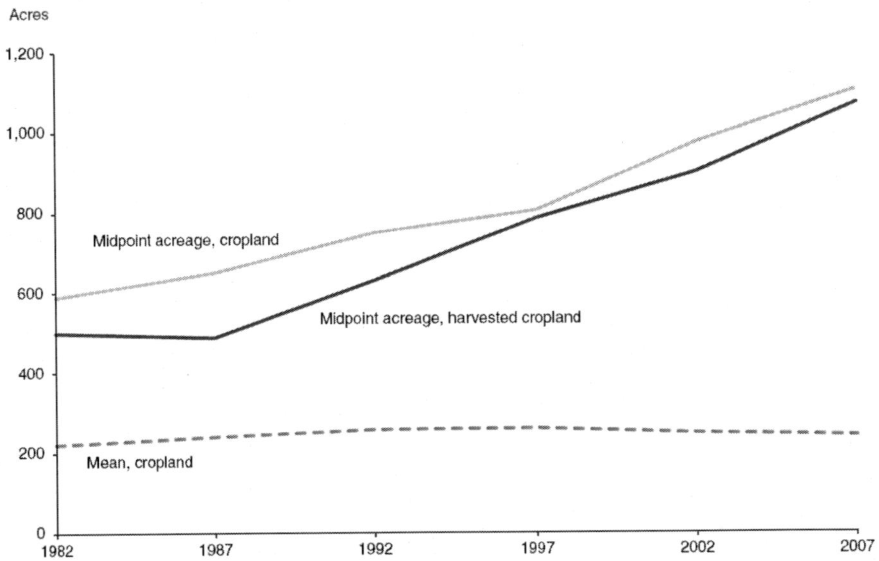

Note: "Midpoint acreage" defined—half of all cropland acres are on farms with more cropland than the midpoint, and half are on farms with less. "Mean" defined—total number of cropland acres is divided by the total number of farms with cropland.

Sources: Mean cropland is from the Census of Agriculture, while the midpoint acreages are ERS calculations from unpublished census of agriculture data

Figure 3. Three measures of average farm size.

The patterns of cropland consolidation do not rule out a role for organization, technology, or policy factors, and do not explain why consolidation is occurring, but the data suggest that the pace of cropland consolidation may vary with the presence of land and population attributes that constrain farm size.

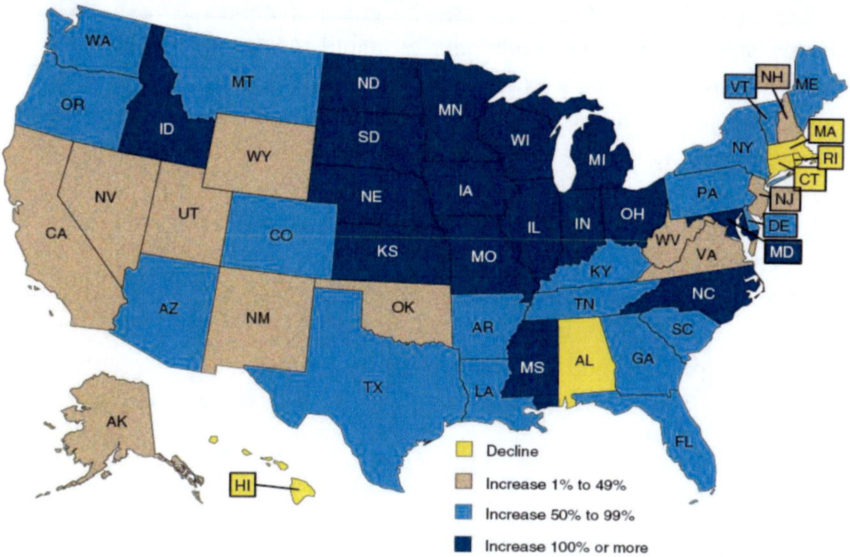

Note: Estimates are "midpoint acreages," for which half of all cropland acres are on farms with more cropland than the midpoint, and half are on farms with less.

Figure 4. Changes in midpoint acreage for cropland, by State, 1982-2007.

Consolidation Among Specific Commodities

A closer look at consolidation among specific crops provides further insight into the trends observed in total cropland. Table 1 summarizes data on harvested acres for major commodities in 2007 and crop receipts for 2007, 1982, and 1950 and reveals three key details. First, four crops (corn, hay, soybeans, and wheat) accounted for over 83 percent of harvested crop acres in 2007. Developments in these few crops drive national trends in midpoint acreages for all cropland combined.

Accounting for Cross-State Differences in Consolidation

ERS researchers related cross-State differences in cropland consolidation to four attributes of a State's agriculture in 1982, the beginning of the period: the share of a State's cropland that was harvested, the share of harvested cropland that was in hay, cropland density (cropland as a share of land area), and population density (population per square mile).

The two density measures are weighted averages across counties within a State, where the weights are cropland. All of the data were from the Census of Agriculture except for population density, which is drawn from Census Bureau (U.S. Department of Commerce) data. Density measures are weighted by cropland to measure density in those parts of a State where the cropland is located. For example, high population density in Manhattan (New York County) is irrelevant because none of New York State's cropland is in that county.

Consolidation was measured as the growth in a State's midpoint acreage for harvested cropland (the logarithm of the ratio of the 2007 to the 1982 midpoints), and consolidation was the dependent variable in an ordinary least squares regression with the four attributes as explanatory variables:

Variable	Mean	Standard deviation	Coefficient	t-statistic	Standardized coefficient
Intercept	-	-	-0.1997	0.87	
% harvested	71.5	12.4	0.0111	3.23	0.138
% hay	28.5	22.9	-0.0034	1.99	0.078
Population density	141.7	183.7	-0.0008	4.12	0.147
Cropland density	35.0	22.3	0.0041	2.16	0.091

Notes: The mean of the dependent variable was 0.51. The H^2 coefficient was 0.64. The standardized coefficient is the coefficient times the standard deviation (impact on the dependent variable of a 1-standard deviation change in an explanatory variable).

This simple model accounts for 64 percent of the cross-State variation in the growth in farm size. Growth in midpoint acreage is noticeably higher in States with high cropland density, low population density, high shares of harvested cropland, and low shares of hay. Each coefficient is statistically and substantively significant, and each explanatory variable shows substantial cross-State variation, to judge from their standard deviations.

Specifically, a one standard deviation reduction in population density is associated with a 0.147 log point increase in the growth rate of the weighted median—see the last column in the table for standardized coefficients. A one standard deviation increase in the share of cropland that is harvested (12.4 percentage points) is associated with a similar (0.138 log points) increase in consolidation.

Cropland density and hay have smaller but still meaningful impacts. Relative to the sample mean (66.5 percent growth), a one standard deviation change toward slower growth in all four variables would reduce predicted growth to 6.6 percent, while a one standard deviation change to greater growth in all four would raise predicted growth to 162.2 percent. That range encompasses the growth rates of 42 of the 50 States in figure 4.

Table 1. Acreage and revenues, major crops, 1950-2007

Commodity	2007 Harvested acres		Percentage of all crop revenues		
	Acres (millions)	*Percentage*	2007	1982	1950
Field crops	**299.7**	**96.4**	**63.2**	**73.8**	**74.1**
Barley	3.3	1.0	0.5	1.1	1.9
Corn	86.3	27.7	22.7	17.7	9.2
Cotton	10.5	3.4	4.3	6.2	23.1
Hay	58.1	18.7	4.0	2.8	2.9
Oats	1.5	0.5	0.1	0.5	2.1
Rice	2.8	0.9	1.4	2.1	1.5
Sorghum	6.7	2.1	0.9	2.1	1.2
Soybeans	63.9	20.6	15.4	17.3	5.4
Tobacco	0.4	0.1	0.9	4.6	8.6
Wheat	50.9	16.4	7.6	13.6	14.0
Other field crops	15.5	5.0	5.4	5.8	8.3
High-value crops	**11.1**	**3.6**	**36.8**	**26.2**	**25.9**
Vegetables, melons	5.8	1.9	12.9	11.2	11.6
Fruits, nuts, berries	5.3	1.7	12.4	9.4	9.6
Greenhouse/nursery	na	na	11.5	5.6	4.7
All crops	**310.8**	**100.0**	**100.0**	**100.0**	**100.0**

Note: The corn and sorghum estimates exclude crops raised for silage, which are reported in "other field crops." The vegetable acreage estimates include harvested acreage of potatoes and dry beans, and field crops acreage estimates exclude them, to make them consistent with the cash receipts estimates.

Sources: Harvested acres are from USDA, National Agricultural Statistics Service, Census of Agriculture. Cash receipts are from Economic Research Service, www.ers.usda.gov/data/FarmIncome/finfidmu.htm.

Second, harvested acres do not correspond closely to cash receipts for crops. The three high-value categories—vegetables and melons; fruits, nuts, and berries; and greenhouse/nursery crops—accounted for nearly 37 percent of all cash receipts from crops in 2007 but less than 4 percent of harvested acreage. Producers of these commodities use labor and physical capital intensively to produce high sales per acre. Because acreage and receipts do not correspond closely, changes in farm size for high-value crops must be tracked separately, since their developments will be masked in aggregated acreage measures.

Third, long-term shifts have occurred in the mix of crops produced. Cotton, tobacco, and oats accounted for a third of all crop cash receipts in 1950. Those shares dropped sharply between 1950 and 1982, and by 2007, the three crops accounted for just 5 percent of cash receipts. The share held by wheat also declined, as the combined share of corn and soybeans rose by nearly 25 percentage points, and the shares held by the high-value commodity categories rose by 11 percentage points.[5]

ERS researchers measured midpoint acreages for specific commodities and found that the shift to larger farms occurred in all major field crops and in most fruit and vegetable crops. These measures represent enterprises, the part of the farm producing a specific commodity. For example, a farm producing corn and soybeans has a corn enterprise and a soybean enterprise, and the midpoint acreage for corn reflects harvested corn acreage only, not harvested acreage of all crops on farms producing corn.

Field crop estimates are presented in table 2. The midpoint acreage for corn tripled from 200 acres in 1987 to 600 in 2007, a jump that was no doubt affected by the relatively high corn prices of 2007 (which led farms to increase corn acres at the expense of other field crops in 2007). But other crops also showed large long-term increases: midpoint acreages for cotton, rice, soybeans, and wheat all more than doubled between 1987 and 2007.

The exception is hay's midpoint acreage, which increased from 120 acres in 1987 to 160 acres in 2007, a much smaller increase than other field crops. Hay accounted for nearly 60 million acres, or 15 percent of all cropland. As noted above, States with substantial fractions of cropland in hay had less consolidation in all cropland.

Table 3 reports midpoint acreages for 20 vegetable commodities, while table 4 covers 19 fruit, tree nut, and berry crops. The commodities listed in the two tables represent 92 percent of all vegetable, fruit, tree nut, and berry acres in the 2007 Census of Agriculture.[6] A wider range of shifts is reported here—midpoint acreages for plums and cantaloupes declined between 1987 and 2007, and a few other commodities showed little change—but most showed large increases. Of the 39 commodities listed in tables 3 and 4, 30 showed increases of at least 50 percent between 1987 and 2007. The average increase across commodities was 107 percent.[7] All commodities listed in tables 3 and 4 showed increased midpoint acreages between 1987 and 1997, even if a few are small increases. In 1997-2007, 31 of 39 commodities showed increases, and 12 of those increased by more in 1997- 2007 than in 1987-1997.

Table 2. Acreage shifted to larger field crop farms, 1987-2007

Commodity (field crops)	1987	1997	2007
	Midpoint acreage, harvested acres		
Corn	200	350	600
Cotton	450	800	1090
Rice	295	494	700
Soybeans	243	380	490
Wheat	404	693	910
Hay	120	140	160

Note: Midpoint acreages are the enterprise farm size, in harvested acres, at which half of all harvested acres are on larger enterprises, and half are on smaller enterprises.

Source: ERS calculations from unpublished Census of Agriculture records, 1987, 1997, 2007.

Table 3. Acreage shifted to larger vegetable and melon enterprises, 1987-2007

Crop	1987	1997	2007
	Midpoint acreage, harvested acres		
Asparagus	160	200	240
Beans, snap	221	225	323
Broccoli	440	780	1,000
Cabbage	113	162	300
Cantaloupe	400	431	388
Carrots	350	900	600
Cauliflower	240	400	400
Cucumber	115	250	505
Lettuce	949	1,461	1,815
Onions, dry	115	220	320
Peas	100	125	179
Peppers, bell	88	180	300
Potatoes	350	556	990
Pumpkins	20	24	30
Spinach	162	242	423
Squash	35	60	72

Table 3. (Continued)

Crop	1987	1997	2007
	Midpoint acreage, harvested acres		
Sweet corn	100	173	250
Sweet potatoes	140	250	474
Tomatoes	400	589	820
Watermelons	80	100	150

Note: Midpoint acreages are the enterprise farm size, in harvested acres, at which half of all harvested acres are on larger enterprises, and half are on smaller enterprises.

Source: ERS calculations from unpublished Census of Agriculture records, 1987, 1997, 2007.

Table 4. Acreage shifted to larger fruit and nut enterprises, 1987-2007

Crop	1987	1997	2007
Non-citrus fruits	*Midpoint acreage, harvested acres*		
Apples	83	122	146
Avocadoes	40	54	40
Cherries, sweet	32	40	65
Cherries, tart	65	89	150
Grapes	205	306	320
Nectarines	70	120	186
Peaches	92	100	120
Plums; prunes	179	250	160
Pears	50	66	75
Citrus fruits			
Grapefruit	320	478	556
Lemons	176	177	176
Oranges	450	769	1,113
Tree nuts			
Almonds	203	292	450
Pecans	102	125	117
Pistachios	465	627	627
Walnuts	85	126	172
Berries			
Blueberries, tame	50	54	75
Cranberries	90	96	99
Strawberries	24	60	120

Note: Midpoint acreages are the enterprise farm size, in harvested acres, at which half of all harvested acres are on larger enterprises, and half are on smaller enterprises.

Source: ERS calculations from unpublished Census of Agriculture records, 1987, 1997, 2007.

To sum up, cropland consolidated onto larger farms in the last three decades. The shifts mirror similar shifts in farm size occurring in livestock. (See box: "Consolidation in Livestock Production.") Consolidation has been persistent, in that it has continued throughout the 1982-2007 period covered by the census data. It has been widespread across States and crop enterprises. Finally, the shifts have also been large—centering on a doubling of midpoint farm size over 20-25 years for national, State and commodity estimates.

Consolidation in Livestock Production

ERS researchers measured livestock consolidation with midpoint sizes based on herd inventory for dairy cows and annual sales or removals for other livestock. In 1987, the midpoint dairy herd size was 80 cows; by 2007, it was 570 cows. The change in hogs was even more striking, from 1,200 hogs removed in a year to 30,000. But consolidation was widespread: midpoint head sold for fed cattle doubled between 1987 and 2007, while those for broilers and cow-calf operations (cattle, less than 500 pounds) more than doubled.

Technology, in the form of economies of scale, played an important role in shifts to larger livestock operations, particularly in hog, dairy, fed cattle, and poultry production (MacDonald and McBride, 2009). Hog, poultry, and fed cattle production also became much more tightly integrated with processors over time, and changes in processor location, size, and contracting practices may have played a further role in livestock consolidation.

	Midpoints		
Commodity	1987	1997	2007
	Annual head removed or sold		
Livestock			
Broilers	300,000	480,000	681,600
Hogs	1,200	11,000	30,000
Fattened cattle	17,532	38,000	35,000
Cattle, <500 lbs.	50	65	128
		Milk cow inventory	
Dairy	80	140	570

Note: The midpoint is defined as the enterprise size, in number of head, at which half of animals are on larger enterprises and half are on smaller enterprises.
Source: ERS calculations from unpublished Census of Agriculture records, 1987, 1997, 2007.

Continuing Advantages to Size in Crop Farming

Larger crop farms perform better financially, on average, than smaller farms. The differences reflect lower costs per unit of production and not higher revenue. In turn, larger farms appear to be able to realize more production per unit of labor and capital. These financial advantages have persisted over time, which suggests that shifts of production to larger crop farms will likely continue in the future.

Larger farms realize higher rates of return on equity (table 5).[8] For each crop, profit measures generally increase from one acreage class to the next, and the differences are large in total. The estimates are drawn from 2008-2011 data, but the pattern holds for other years as well. These differences in profit performance do not stem from differences in revenue—larger grain and oilseed farms realize the same value of production per acre as smaller farms. The smallest vegetable farms (less than 50 acres) realize a higher value of production per acre than larger farms, but otherwise no apparent relation exists between size and revenue per acre at fruit and vegetable farms. These advantages must, therefore, arise from differences in farm production costs.

Labor hours per harvested acre decline sharply as harvested acreage increases for corn, wheat, and soybean farms (table 6).[9] Farms harvesting more than 2,000 acres use less than half as much labor per acre as farms harvesting fewer than 500 acres. Table 6 also reports the value of equipment and structures (assets) owned by the farm, per harvested acre. Assets per acre decline persistently as farm size increases; they are 35 to 50 percent lower for the largest farms than for those in the 250-499 acre class.

Table 7 reports the same measures for fruit and vegetable operations. Higher labor and capital use per acre in fruits and vegetables than in the field crops reflects the capital-intensive nature of fruit and vegetable production. However, the same general size relationship holds: labor hours per harvested acre decline as acreage increases, and capital per acre (the value of equipment and structures) also declines, except for labor for the largest size class (1,000 or more acres) among fruit operations. Moreover, the differences are large enough to be financially meaningful. In each commodity class, larger operations appear to be able to apply their labor and capital to more acres than smaller farms; with regard to capital, that suggests that larger farms get more hours of use in a year. Of course, this pattern may also reflect a more subtle relationship—that farms that are able to use labor and capital more intensively may also be better able to expand.

Leasing, Contract Labor, and Custom Services on Crop Farms

Tables 6 and 7 report on the labor that farming operations directly employ and the capital equipment that the operations own. But farms may access labor and capital in other ways:

Table 5. Larger crop farms realize higher average profit rates

	Farm specialization				
Harvested acres	Corn	Soybeans	Wheat	Fruits/ nuts	Vegetables/ melons
	Class-average rates of return on equity, 2008-2011				
Less than 10	-	-	-	-1.4	-0.9
10-49	-	-	-	-0.2	2.7
50-99	-	-	-	4.1	-1.9
Less than 100	-0.9	-1.3	-2.6	-	-
100-249	1.2	-0.03	-0.6	5.4	8.1
250-499	2.9	1.1	0.6	3.8	17.8
500-999	4.8	1.7	0.4	7.1	8.9
1,000 or more	-	-	-	10.7	17.9
1,000-1,999	5.3	5.4	4.6	-	-
2,000 or more	8.0	8.2	5.5	-	-

Notes:

[1] The sample consists of farms with no livestock production (crops only), and primary specializations in one of the five commodity classifications noted above. The primary specialization is the commodity accounting for the largest share of farm sales. Farms with livestock were omitted in order to simplify the analysis and focus on financial performance for crops.

[2] The numerator of the rate of return on equity is net farm income, minus charges for unpaid operator labor and management contributions, while the denominator is net worth (assets minus debt). The table presents class averages—total returns across all farms in a commodity/acreage class, divided by total net worth in the class.

[3] Pooling observations across years provides a larger sample size and more reliable estimates.

Source: Agricultural Resource Management Survey, Phase III, all versions, 2008-2011 pooled (17,351 observations: 6,619 in corn; 2,810 in soybeans; 2,079 in wheat; 4,492 in fruits; and 1,351 in vegetables).

- By leasing capital equipment and structures from dealers or from other businesses that rent out equipment and structures. Such assets are not recorded as assets of the farm business.

- By contracting for workers who are hired, employed, and paid through a contract labor provider.
- By contracting with custom service providers that bring their own labor, and may bring their own equipment, to perform field tasks such as spraying or harvesting.

These transactions afford several advantages, including flexibility to change the size of operations, avoidance of the financial risks of investment in expensive capital equipment, and the freedom to focus on specific farm tasks, while relying on custom providers' expertise for other tasks. If larger farmers use these services more than small farms, then an exclusive focus on capital assets and labor provided by the farm may understate the costs of larger operations and overstate their cost advantages over smaller operations.

Table 6. Labor and capital on corn, soybean, and wheat farms

	Harvested acres					
	Less than 100	100-249	250-499	500-999	1,000-1,999	2,000 or more
	Mean hours per harvested acre					
Labor (all)						
Corn	38.6	12.3	7.8	5.7	3.5	2.7
Soybeans	45.7	10.4	7.3	5.8	3.8	3.0
Wheat	40.4	8.7	5.8	5.3	3.2	2.2
	Hired labor hours as a percent of total labor hours					
Hired labor						
Corn	5.0	2.9	4.6	10.2	16.9	31.2
Soybeans	2.7	5.2	7.4	14.6	16.4	36.0
Wheat	4.0	3.2	3.4	16.3	19.5	20.5
	Equipment and structures assets ($) per harvested acre					
Capital						
Corn	2,532	847	683	568	505	43
Soybeans	2,880	826	640	535	387	33
Wheat	3,325	588	396	320	278	24

Note: The labor and hired labor panels are based on version 1 only (2,087 observations), while the capital measures are drawn from all versions (11,508 observations).

Source: Agricultural Resource Management Survey, Phase III, 2008-2011 pooled.

Custom services are used extensively by many corn, soybean, and wheat operations, but (except in wheat) no apparent relationship exists between farm size and the likelihood that a farm will use custom services (table 8). However, larger farms are much more likely to use contract labor or to lease capital equipment and structures. This suggests that table 6 may understate the farm use of labor and capital when all sources are accounted for, and that it may therefore overstate the advantages of large farms regarding input use.

However, the bottom panel of table 8 shows that even though larger farms are more likely to incur expenses for each practice, dollar expenses fall sharply, per acre, as acreage increases. If there is a fixed set-up cost for these items, as well as a component of expenses that varies with acreage, then larger farms could have lower expenses per acre because they are able to apply the contracted inputs and custom services that they hire to more acres per hour of hire. Regarding the intensity of resource use, table 8 reinforces the message from table 6 that larger crop farms appear to be able to apply a given amount of capital and labor to more acres.

Producers of fruits and vegetables make extensive use of contract labor, leased capital, and custom services (table 9). In general larger farms are considerably more likely than smaller farms to use each of these practices, but there is no apparent relationship between farm size and the per acre expenses that are incurred for these services. Even when contracted and leased inputs are considered, larger fruit and vegetable farms are able to apply a given amount of labor and capital across more acres, on average, realizing lower costs per acre.

Table 7. Labor and capital on fruit, tree nut, and vegetable farms

	Harvested acres of fruits, tree nuts, and vegetables						
	Less than 10	10-49	50-99	100-250	250-499	500-999	1,000 or more
	Mean hours per harvested acre						
Labor							
Fruit/nuts	564.7	202.9	145.5	145.1	128.4	74.2	159.6
Vegetables/melons	849.3	204.4	150.9	98.5	59.9	46.4	32.8
	Hired labor hours as a percent of total labor hours						
Hired labor							
Fruit/nuts	7.4	23.6	43.0	52.1	51.2	58.2	55.6
Vegetables/melons	5.1	14.2	37.7	41.7	43.4	47.6	61.4

Table 7. (Continued)

	Harvested acres of fruits, tree nuts, and vegetables						
	Less than 10	10-49	50-99	100-250	250-499	500-999	1,000 or more
	Equipment and structures assets ($) per harvested acre						
Capital							
Fruit/nuts	15,862	5,154	3,280	2,173	1,693	1,314	971
Vegetables/ melons	26,439	4,677	2,531	2,240	1,469	1,188	885

Note: The labor and hired labor panels are based on version 1 only (1,471 observations), while the capital measures are drawn from all versions (5,843 observations).

Source: Agricultural Resource Management Survey, Phase III, 2008-2011 pooled.

The financial estimates show continuing advantage to large farms, which should favor further consolidation. However, the pace of cropland consolidation also appears to have slowed in recent years. The midpoint acreage for cropland, measured with data from the Census of Agriculture, nearly doubled between 1982 and 2007, to 1,105 acres (fig. 3). Data from the 2012 Census have not yet been released, but the midpoint acreage estimated from the 2011 ARMS (see fig. 1) was 1,100 acres, just 1 percent greater than the ARMS-based estimate of 1,093 acres in 2007 (in fig. 3), and slightly below the 2007 Census estimate.[10] More definitive evidence will be available in data from the forthcoming 2012 Census of Agriculture.

Some recent developments in consumer demand may favor smaller family farms. For example, consumer and retailer interest in local food production is growing (Low and Vogel, 2011). Farms that produce for local markets are considerably smaller, on average, than other farms. In 2010, the midpoint acreage among crop farms that produce for local markets was 310 acres, compared to 1,100 acres for other crop farms. Among fruit and vegetable producers, the midpoint acreage was 168 acres, compared to 675 acres among other fruit and vegetable farms. Thus, shifts of consumption to local markets, and hence to the farms that supply them, do appear to favor smaller operations.[11] However, farms that sell to local markets account for less than 5 percent of U.S. cropland.

Table 8. Use of contract labor, custom work, and leased capital in grain production

	Harvested acres					
Practice and commodity	Less than 100	100-249	250-499	500-999	1,000-1,999	2,000 or more
	Percent of farms using practice					
Custom work						
Corn	49.6	51.8	48.3	49.0	46.7	48.7
Soybeans	49.9	46.3	40.3	44.8	41.9	45.8
Wheat	31.3	40.7	45.7	49.9	52.8	60.0
Contract labor						
Corn	2.9	3.0	6.7	6.8	13.1	11.8
Soybean	1.9	3.2	4.5	5.7	9.8	17.5
Wheat	12.8	11.7	9.4	10.3	14.1	14.0
Leased capital						
Corn	7.9	7.2	14.6	17.5	23.4	37.9
Soybeans	4.6	12.1	12.7	12.8	12.6	23.9
Wheat	6.9	10.1	12.8	18.8	17.2	28.6
	Expense ($) per harvested acre for all three practices					
Expense						
Corn	31.13	21.76	15.10	12.51	12.07	10.86
Soybeans	25.57	18.90	13.58	11.23	7.05	10.56
Wheat	26.53	17.90	14.63	11.90	11.40	9.98

Notes: A farm is considered to "use a practice" if it has positive expenses for the practice, while expense per acre is the sum of expenses for custom work, contract labor, and leased capital.

Source: Agricultural Resource Management Survey, Phase III, all versions, 2008-2011 pooled.

Prices for farm products and for energy inputs were extraordinarily volatile in recent years (Baffes, 2013). In turn, that volatility increased the financial risks faced by farmers and by lenders. It also created uncertainty about the future path of Government policy, as policymakers and commodity groups debated the future orientation of crop insurance, commodity programs, and biofuels programs. The risk and uncertainty of the last several years may have slowed cropland consolidation, as farm operators and lenders pause to better understand the new financial climate.

Table 9. Use of contract labor, custom work, and leased capital in fruits and vegetables

Practice and commodity	*Harvested acres*						
	Less than 10	10-49	50-99	100-249	250-499	500-999	1,000 or more
	Percent of farms using a practice						
Custom work							
Fruits/nuts	25.8	41.7	56.5	56.7	63.1	70.3	71.6
Vegetables/melons	10.9	16.0	9.2	31.1	42.2	54.5	64.7
Contract labor							
Fruits/nuts	31.3	47.7	52.0	59.4	64.8	70.3	79.5
Vegetables/melons	3.1	13.8	17.6	29.0	36.0	33.3	56.3
Leased capital							
Fruits/nuts	7.2	9.8	21.3	22.4	30.6	39.0	44.6
Vegetables/melons	3.7	4.4	4.9	25.5	43.2	39.9	59.2
	Expense ($) per harvested acre for all three practices						
Expense							
Fruits/nuts	526	550	625	507	557	631	475
Vegetables/melons	234	122	83	119	203	202	239

Notes: A farm is considered to "use a practice" if it has positive expenses for the practice, while expense per acre is the sum of expenses for custom work, contract labor, and leased capital.

Source: Agricultural Resource Management Survey, Phase III all versions, 2008-2011 pooled.

DRIVERS OF CONSOLIDATION

ERS researchers focused on three factors:

- **Technology**. Labor-saving innovations—in equipment, chemicals, seeds, tillage practices, and information technology—allow farmers to operate larger farms.
- **Changes in the organization of farms**. Over a long period, crop production has separated from livestock production as individual farms have focused on a few commodities and has concentrated in regions suited for adopting scale-increasing technology. Specialization and investment in costly capital equipment can be risky

endeavors, and farmers also adopted methods of financing and transacting to reduce risks.

- **Government policy**. Federal initiatives affect farm structure through many channels—some lead to larger farms, while others support smaller farms. Some programs also affect the risks perceived by producers, and they can indirectly affect farm size by the way farmers adapt to these risks.

Drivers of Consolidation: Technology

Economists studying shifts in farm size often focus on the role of technology, and in particular on two concepts: scale economies and labor-saving technological change. Scale economies are said to exist if expansions of output lead to reductions in per-unit costs of production holding input prices constant.[12] Scale economies are technologically based, in that they arise from savings in input use, and not because a larger firm realizes lower prices for inputs.[13] They may be driven by certain fundamental physical relationships in production and may arise from the higher specialization in tasks that increased production can offer. They often arise because larger production runs may allow for the use of capital equipment, and more capital-intensive production, that would not be viable for smaller production runs.

Economies of scale matter for firm and plant size in non-agricultural industries. They also appear to matter, over certain ranges of production, in agriculture and particularly in livestock production (MacDonald and McBride, 2009; Miller et al., 1981). Nonetheless, most economists are skeptical that scale economies usefully explain increased farm sizes. Most farms remain fairly small, compared to firms in other industries, and crop production still covers a wide range of viable farm sizes. Moreover, because many large pieces of capital equipment are mobile and can be rented out to small operations for modest production runs, multiple small farms can effectively share equipment.

Instead of focusing on scale economies, analysts more often focus on labor-saving innovations, which are especially key for agriculture, where most farms are family-operated, with a certain amount of labor available for operation of the farm (Kislev and Peterson, 1982; Gardner, 2002; Cochrane, 1993). Innovations that reduce the amount of labor required for field operations allow farming of more acres. Likewise, innovations that give farmers more accurate or timely information often increase the amount of land they can effectively manage. Labor-saving innovations do not have to explicitly favor larger farms to affect farm structure; instead all they have to do is facilitate the expansion of farms.

Equipment

The tractor remains a primary example of a labor-saving innovation. The first general-purpose tractor capable of operating among field crops was introduced in 1924, and its use spread until around 1960, with tractors steadily replacing horses and mules in field tasks. Olmstead and Rhode (2001) estimate that by 1960 the tractor had replaced 23 million draft animals, and the 79 million acres of land used to grow feed for them were reallocated to other uses. The tractor reduced the amount of labor time required to perform field operations, and required less labor time to maintain than draft animals. Olmstead and Rhode estimate that the tractor reduced the labor required to produce the agricultural output of 1960 by 1.7 million workers, about 24 percent of farm employment in that year. Finally, they estimate that tractor adoption increased average farm size by 58 acres, about 37 percent of the growth in average farm size between 1910 and 1960, as it allowed a single farmer to operate more acres.

The tractor wasn't the only labor-saving mechanical innovation introduced during the period. Mechanical harvesters, sprayers, and planters, designed for many specific crops, further reduced labor requirements (Street, 1957; Rasmussen, 1962, 1968; Reimund et al., 1981; Whatley, 1983; Calvin and Martin, 2010). In each case, total labor requirements were reduced, leading to reductions in the total farm workforce, but families that remained in agriculture could expand the farm operation with the labor and capital equipment available to them. In some cases, mechanical innovations required complementary biological innovations that created crops that could be more easily harvested mechanically.

Since the time covered by Olmstead and Rhode, tractors and other farm equipment have continued to get larger and faster, allowing individual farmers to manage yet more acreage. Bechdol et al. (2010) argue that a farmer could plant 40 acres of field crops in a day using the tractors and planters available in 1970, but 420 acres in 2005, and 945 in 2010, with larger and faster tractors hauling planters that covered far more rows than the 1970 version (table 10). For similar reasons, a farmer could harvest more than 12 times as much in a day in 2010 as in 1970.

Moving larger pieces of equipment between fields takes more time and expense, and larger pieces also require more setup time at fields. For these reasons, bigger and faster equipment is most valuable where fields are large, flat, and contiguous. Therefore, it is more valuable, and has greater effects on farm size, in the Western Corn Belt, the Plains, and the Delta, than in the Eastern Corn Belt, Appalachia, and the Northeast.

To see this effect, consider data on recent purchases of very large tractors—500 horsepower or more (which sell for an average price of over $250,000).[14] Between 2005 and 2011, 60 percent of those tractors were sold in just five Corn Belt and Plains States—Illinois, Indiana, Iowa, Minnesota, and North Dakota (table 11). Those are major agricultural States, but collectively account for just 27 percent of U.S. cropland, far below their share of large tractor sales. The next five most important States for large tractor sales (South Dakota, Texas, California, Michigan, and Ohio) accounted for 17 percent of sales and 20 percent of cropland, while all other States account for 53 percent of cropland and just 23 percent of large tractor sales.

Table 10. Changes in planting and harvesting machinery in field crops, 1970-2010

Year	Planting efficiency		Harvesting efficiency	
	Technology	*Outcome*	*Technology*	*Outcome*
1970	4 rows @ 2 mph	40 acres/day	4 rows, 12 hrs/day	4,000 bu./day
2005	16 rows @ 6 mph	420 acres/day	12 rows, 12 hrs/day	30,000 bu./day
2010	36 rows @ 6 mph	945 acres/day	16 rows, 12 hrs/day	50,000 bu./day

Notes: mph=miles per hour; hrs=hours; bu=bushels
Source: Bechdol, Gray, and Gloy (2010)

Table 11. Sales of large (500 horsepower or more) four-wheel drive tractors, by State, 2005-2011

State	Large tractor sales, 2005-2011		Cropland acres	Cropland density	Share of cropland non-harvest hay	
	Units	*Million $*	*Millions*	*Percent*	*Percent*	*Percent*
Top five, sales	3,240	837.7	111.6	71	11	7
Minnesota	1,119	287.2	21.9	69	12	9
North Dakota	864	223.0	27.5	69	20	11
Illinois	633	166.6	23.7	73	5	3
Iowa	381	97.8	26.3	77	10	5
Indiana	243	63.2	12.1	64	5	5
2nd five, sales	893	229.1	80.9	43	27	22

Table 11. (Continued)

State	Large tractor sales, 2005-2011		Cropland acres	Cropland density	Share of cropland non-harvest hay	
	Units	*Million $*	*Millions*	*Percent*	*Percent*	*Percent*
South Dakota	224	57.5	19.1	53	20	21
Texas	191	49.2	33.7	38	43	27
California	183	45.4	9.5	27	19	23
Michigan	160	42.2	7.8	43	12	17
Ohio	135	34.6	10.8	59	8	12
Other States	1,246	317.0	214.0	40	29	29
United States	5,379	1,383.8	406.6	49	24	21

Note: Cropland contiguity is the weighted average share of land in cropland, where the observations are counties and the weights are cropland.

Sources: Tractor sales are from Agricultural Equipment Manufacturers data provided to ERS, while cropland data are from Census of Agriculture.

Land attributes in the five leading States are different from other States. Much higher shares of cropland in those States are harvested than in other States, and much lower shares are planted to hay, which is frequently grown on hillier and more marginal land. Moreover, cropland density is much higher in the top five States (71 percent of the land area is in cropland, on average, compared to 43 percent in the next five States and 40 percent for the rest of the country).[15] High cropland density indicates large and contiguous fields, where larger and faster tractors will be most effective.

Chemical Pesticides

Pesticides are substances used to control pests, including herbicides for weed control, insecticides to control insects, fungicides to control fungi and other disease pathogens, nematocides to control parasitic worms, and rodenticides for rodents. Pesticides can be synthetic or natural, and they are not the only means by which farmers might control pests. Farmers can control weeds through weeding, either by hand or with mechanical tillage, and they can control pests with management practices like crop rotations and interplanting.[16]

The use of chemical herbicides expanded between 1960 and 1980, from 35 million pounds of active ingredients to 469 million, as the proportion of acres treated with herbicides rose to over 90 percent for corn, cotton, and soybeans, and nearly 50 percent for wheat and potatoes (Fernandez-Cornejo, et al., 2013). Fernandez-Cornejo and Pho (2002) show that herbicide prices fell sharply compared to labor and machinery prices in this period, leading to two types of adjustments. First, farmers substituted existing chemical herbicides for other labor-intensive methods of weed control. Second, the relatively high prices for labor and machinery led to more research and development that would generate improvements in herbicides and other chemicals, leading to further substitution of chemicals for labor.

After 1980, the total pounds of herbicides applied stabilized and even declined for some crops, as the quality of herbicides improved. Newer substances more effectively controlled weeds with fewer applications, less toxicity, and shorter persistence in the soil (Fernandez-Cornejo and Pho, 2002; Fernandez-Cornejo et al., 2013). When adjusted for improved herbicide quality, prices of chemical methods of weed control continued to fall relative to labor and machinery prices after 1980, and farmers shifted to newer labor-saving herbicides, even as total herbicide applications stopped rising.

In their study using annual time series data for the United States covering 1948-1995, FernandezCornejo and Pho (2002) find a strong labor-saving bias to the adoption of herbicides. They estimated that a 1-percent increase in the price of labor relative to herbicides would lead to a 0.23 percent increase in the ratio of herbicides to labor used in production, during the year of the increase. The long-run effect—covering longer term input substitution and the effects of induced innovation—was quite large; a 1-percent increase in the relative price of labor would lead to a 13.5-percent increase in the ratio of herbicides to labor use.

Substituting herbicides for labor increased the amount of cropland that a farm family could manage by reducing the labor and management time required per planted acre of a given crop. The impacts were likely sizable between 1960 and 1980, when herbicide use spread widely. Later herbicide improvements likely allowed for further incremental expansions in farm size, intertwined with changes in crop genetics and tillage practices.

Genetically Engineered Seeds

Developments in seed genetics have also saved labor, either directly or in combination with other innovations (Gardner, 2002; Cochrane, 1993). For example, mechanical harvesters for processing vegetables, which greatly reduced labor hours, necessitated the complementary development of crop products that could withstand mechanical handling in harvest (Rasmussen, 1968; Schmitz and Seckler, 1970; Reimund et al., 1981).

More recently, the spread of genetically engineered (GE) seeds in corn, cotton, and soybeans (fig. 5) has affected farm production practices and the allocation of operators' time.[17] Herbicide-tolerant (HT) seeds, used in all three crops, allow farmers to apply one herbicide product at a post-emergent stage, thereby replacing several herbicides applied at different times (Fernandez-Cornejo, 2007). Use of HT seeds reduces machine and machine operators' time, as well as the time used to evaluate the need for and plan of weed management. Other GE seeds (Bt), used in corn and cotton, are pest-resistant. In principle, they allow farmers to forego spraying pesticides, thereby saving machine and machine operator hours, although some farmers facing infestations may have simply accepted lower yields and did not use chemical pesticides. To the extent that GE seeds reduce the amount of time needed to plan and operate a given amount of acreage, they potentially free time to manage a larger farm.

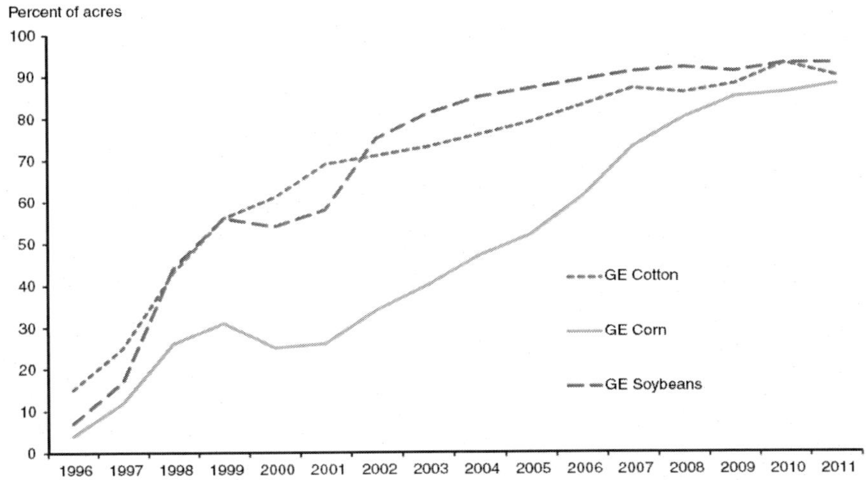

Source: Fernandez-Cornejo (2005) for 1996-1999; USDA NASS, *Acreage*, for 2000-2011.

Figure 5. Adoption of genetically engineered (GE) seeds in the United States.

Gardner et al. (2009) analyzed the linkages between labor hours used per acre and GE seed adoption, using ARMS corn, cotton, and soybean data collected in 2001, 2002, and 2003, respectively.[18] They find that adoption of HT seeds was associated with statistically significant reductions in labor hours in soybean and cotton fields, but not in corn fields, where HT adoption was still quite low in 2001.

ERS researchers used ARMS data for 2006 and 2008 to assess the relationship between GE seed adoption and hours of farm labor for corn, cotton, and soybean farms, while controlling for total cropland and the mix of crops on the farm, the use of capital equipment, and the use of custom services and contract labor, each of which can substitute for on-farm labor. (See box: "GE Seeds and Labor Time.")[19]

Results suggest that farms with acreage in HT crops realized substantial reductions in labor hours, per acre of land. For a farm with 1,500 acres of corn and soybeans, but no land planted to GE seeds, labor use on the farms would amount to 4,421 hours. With all of the land planted to HT traits, and none to Bt, the estimated labor hours would fall to 3,160 hours, freeing 1,261 hours for off-farm work, family time, or expanded farm production. Most farms don't face such stark choices between all or nothing: alternatively, the model estimates indicate that labor hours per acre would fall by 15.5 percent on farms with 75 percent of acreage in HT seeds compared to farms with 25 percent in HT. In contrast, the analysis shows no significant association between Bt adoption and labor hours.[20]

If GE seeds save management and labor time, farmers could use the freed labor time to work off the farm or to devote more time to family activities. Earlier research on the topic indicates that use of GE seeds was associated with increased earnings from off-farm employment, for a given size of farm (National Academy of Sciences, 2010). That research, driven by data availability, could access data on off-farm earnings but lacked direct data on labor hours, so researchers could not track whether farmers also used the freed time to expand their farming operation. This analysis described above shows a direct link to on-farm labor hours.

Genetically engineered seeds were commercially introduced in 1995, so they cannot account for changes in farm size before that time, and so far they have been used primarily in corn, cotton, and soybeans, so they cannot account for the ubiquitous increase in farm size among many crops. Their effect on corn, cotton, and soybean farms may be limited by the extent of acreage devoted to other crops or by time devoted to livestock production. Nevertheless, GE seeds may partly explain increased consolidation among field crop farms since 1995.

GE Seeds and Labor Time

The 2006 and 2008 ARMS Phase III (Version 1) questionnaires included detailed questions on the use of genetically engineered (GE) seeds and on labor hours for the whole farm. ERS researchers selected a sample of farms from the 2 survey years, consisting of 1,798 operations with corn, cotton, or soybean production but no livestock (most raised other crops as well). Substantial cross-farm differences exist in the share of acreage planted to GE seeds, because of differences in the use of GE seeds for corn, cotton, and soybeans, and because of differences in the acreage of other crops planted on the farm.

ERS researchers sought to determine whether farms with higher shares of acreage planted to GE seeds worked fewer hours on the farm. To that end, a simple ordinary least squares regression was estimated, in which the dependent variable was the logarithm of total hours worked on the farm by all operators and other labor. Researchers controlled for total farmland acreage (in logs), the total dollar value of farm buildings and equipment (also in logs), the share of land that was harvested, dummy variables for the use of contract labor or custom services, a dummy variable to distinguish 2008 from 2006, and measures of the share of production originating in each of 15 field crop categories. The measures of GE acreage were the share of the farm's land planted to herbicide tolerant (HT) and to pest-resistant (Bt) seeds.

Labor hours were higher on farms with more land, more capital equipment, and a higher share of land in harvested acres. They were lower where the farm used contract labor or custom services and were lower in 2008 than in 2006. Hours varied with the mix of crops.

The coefficient on Bt was never statistically significant, but it was consistently positive, in this model and in alternatives analyzed for corn and cotton farms separately. Farmers plant seeds with Bt traits where there is a risk of insect infestation, and those risks may cause the farmer to have to use more hours (per acre) on other control practices. In any case, Bt seeds do not appear to reduce labor hours. In contrast, and consistent with the findings of Gardner, Nehring, and Nelson (2009), HT seeds reduce labor requirements per acre. The coefficient is negative, statistically significant, and large.

Regression results: Labor use on crop farms		
Variable	Coefficient	t-statistic
Intercept	3.533	6.62
Log of land in farm	0.340	8.10
Percent of land harvested	0.519	6.41
Log of physical capital	0.187	4.35
Contract labor used (yes=1)	-0.379	1.43
Custom work used (yes=1)	-0.169	1.97
Bt acreage (percent)	0.167	1.13
HT acreage (percent)	-0.336	3.73
R2	0.38	

Note: Dependent variable: log of annual paid and unpaid labor hours. Other variables include a dummy for 2008, and shares of acreage in each crop (corn, wheat, soybeans, cotton, peanuts, potatoes, barley, oats, tobacco, sorghum, hay, sugar cane and sugar beets, canola, other oilseeds, other crops). Observations are weighted to reflect sample selection probabilities, and standard errors are estimated using the delete-a-group jackknife method to account for complex sample design.

To understand the size of the effect, consider a farm in 2006 with 1,500 acres planted equally to corn and soybeans, with no custom work or contract labor, with $300,000 in equipment. With no land planted to GE seeds, the farm would require an estimated 4,421 hours. With all of the land planted to HT traits, and none to Bt, the estimated labor hours fall to 3,160 hours, freeing 1,261 hours for off-farm work, family time, or expanded farm production. This is an extreme example, given the 100 percent GE assumption, but the impact on hours is striking, as well as the implication that farmers with extensive acreage in HT crops could use those hours to noticeably expand farm size.

HT seeds were approved in 2011 for use in alfalfa, and other GE traits may be available in the future. Like other innovations, they can lead to increased farm size if they are labor-saving. However, other developments may reverse the impact of GE seeds on farm size. In recent years, resistance to glyphosate (the herbicide most commonly associated with HT seeds) has emerged among weeds. If farmers must return to older and more labor-intensive methods of weed control to manage resistance, the effect of GE seeds on farm size may slow and reverse.

Tillage Practices

"No-till" has become an important soil conservation practice in the United States. A no-till system leaves crop residue from the previous harvest on the soil, and soil is left undisturbed from prior harvest to planting, except for the injection of nutrients. By allowing for fewer machinery passes over fields and fewer machine operator hours, no-till saves labor, as well as capital and energy.

Estimates of the share of planted acreage under no-till are reported in table 12. The estimates cover five major field crops for 2000-2010.[21] For each crop shown, no-till shares are higher in later years, although the increases are modest. Horowitz et al. (2010) estimate that shares of planted acreage under no-till expanded at 1.5 percentage points per year between 2000 and 2009, when no-till covered 35 percent of planted acreage for eight major U.S. field crops (barley, corn, cotton, oats, rice, sorghum, soybeans, and wheat). Gardner et al. (2009) find that the use of no-till is associated with large and statistically significant declines in per-acre labor hours in corn, cotton, and soybeans.

HT seeds allow for easier use of no-till practices, and the recent spread of no-till owes something to the expansion of acreage planted to HT seeds. However, these phenomena are not entirely interdependent. While the highest share of no-till acreage is in soybeans, which had the earliest wide application of HT seeds, table 12 shows a rapid expansion of no-till in wheat, which had no HT seeds.

Information Technology

Modern farm equipment embodies another set of innovations known as precision agriculture: management practices and information technologies (IT) that measure and manage intra-field variations in soil attributes, pest presence, product attributes, and production outcomes. Specific technologies include yield monitors that measure and map intra-field variations in yields at harvest; pest monitors that identify the presence of insects and fungi at spraying; GPS systems that map soil, yield, and pest data and transmit the information to vehicles in fields; and variable rate spraying and injection technologies that match the application of chemicals to the intra-field variations in yields, nutrients, and pest presence identified in the other technologies. These technologies, as well as auto-steering and guidance systems for tractors, hold the promise of reducing per-acre use of energy and chemicals primarily, but also of labor and management.

Table 12. Use of no-till on planted cropland, 2000-2010

Crop	2000	2001	2002	2003	2004	2005	2006	2007	2008	2009	2010
	Percent of planted acres in no-till system										
Corn	17	19	-	-	-	24	-	-	-	-	2
Cotton	-	-	-	15	-	-	-	20	-	-	-
Rice	5	-	-	-	-	-	12	-	-	-	-
Soybeans	31	-	35	-	-	-	45	-	-	-	-
Wheat	-	-	-	-	22	-	-	-	-	39	-

Source: Agricultural Resource Management Survey, Phase II, 2000-2010.

In fragile tree fruits like apples and oranges, researchers are experimenting with machines that locate and map fruit using a vision technology, combined with robotic machines that take information from mapping machines and use it to selectively harvest fruit as it matures (Calvin and Martin, 2011). These technologies, which may also apply to other crops, allow farmers to increase the value of production per acre, while reducing labor and management requirements. Other precision technologies, such as laser guidance for field leveling and drip irrigation, enhance water management.

Adoption of precision technologies has been growing: some type of precision technology was used on 58 percent of wheat acres in 2009, up from 14 percent in 1999; on 49 percent of corn acres in 2005, up from 35 percent in 1999; and on 45 percent of soybean acres in 2006, up from 31 percent in 1999 (Schimmelpfennig and Ebel, 2011).

Precision technologies can save on energy, chemicals, and nutrients, and they may also save management time if they allow farmers to gather and retain more accurate information in the course of field operations instead of in separate field visits. Because the innovations are embodied in expensive capital equipment, they also likely create scale economies: that is, given their expense, many of them are likely to only be economical on large operations that can spread their costs over large production volumes or through custom provision of services.

So far, most use of precision technologies has involved yield monitors, and widespread use of those has occurred recently, so precision technologies did not play a major role in the past growth of farm size (Schimmelpfennig and Ebel, 2011). Still, their role is likely to grow in the future if they are labor-saving, and more importantly, if they provide farm managers with localized information that they would otherwise only obtain through time-intensive personal inspections.

Drivers of Consolidation: Farm Organization and Location

During the 20th century, crop production largely separated from livestock production, and crop farmers focused on producing a few commodities. Greater specialization provided a spur to increased size, and crop production also shifted toward regions suited for adopting scale-increasing technology. Specialization, and investment in costly capital equipment, can be risky endeavors, and farmers also adopted new methods of financing and transacting to reduce risks.

Farm Specialization

Most U.S. farms kept a variety of livestock in the first half of the 20th century, and they raised feed crops for their animals as well as cash crops for sale. While they became more specialized during the first half of the century, at least half of U.S. farms still had hogs, milk cows, or chickens in 1960 (fig. 6). The shift away from widespread and diversified livestock production accelerated after that date, and continued through 2000, when less than 10 percent of farms had chickens, milk cows, or hogs, and those that did usually specialized in one species.

The presence of livestock affects farm crop choices. In 1900, most farms grew corn, usually to feed to their own animals (fig. 7). As farmers ceased small-scale diversified livestock production, they also stopped producing feed for their small herds and flocks. By 2010, only one in six farms grew corn, and those farms usually specialized in crops, with no livestock.[22]

Livestock production concentrated onto fewer farms, and those farms focused on livestock, although most continued to grow some crops for sale and for manure utilization (MacDonald and McBride 2009). Today, almost all poultry production, and most pork and fed cattle production, rely exclusively on purchased feeds, while dairy production combines onfarm production with purchased feed. As a result, most crop production is concentrated on farms with no livestock, focusing on growing a few crops. Commercial markets for feed became deeper, with more transactions and greater volumes of feed sales, as less feed was produced for animals on the farm and more was sold by crop farms to intermediaries for processing and sale to livestock farms as feed components or complete mixes. Deeper markets provided more reliable services, a wider range of desirable feed attributes for purchase, and more certain market outlets for farmers.

Crop farmers who no longer pursued small-scale diversified livestock production had more free time available, which could be applied to work off the farm or to expanded crop production on more acres. As a result, changes in farm organization toward greater specialization also provided an impetus for crop farmers who wished to farm full-time to operate more cropland.

Once they left livestock production, crop farmers needed to decide on the mix of crops to produce. If all of a farmer's cropland has common soil and climatic attributes, then a single specific crop might be the most profitable for the farm, for a given combination of product and input prices. More specialized farms can also realize more intensive use of specialized equipment and structures, and greater accumulation of crop-specific knowledge by producers, leading to reduced production costs (Holmes and Lee, 2012).

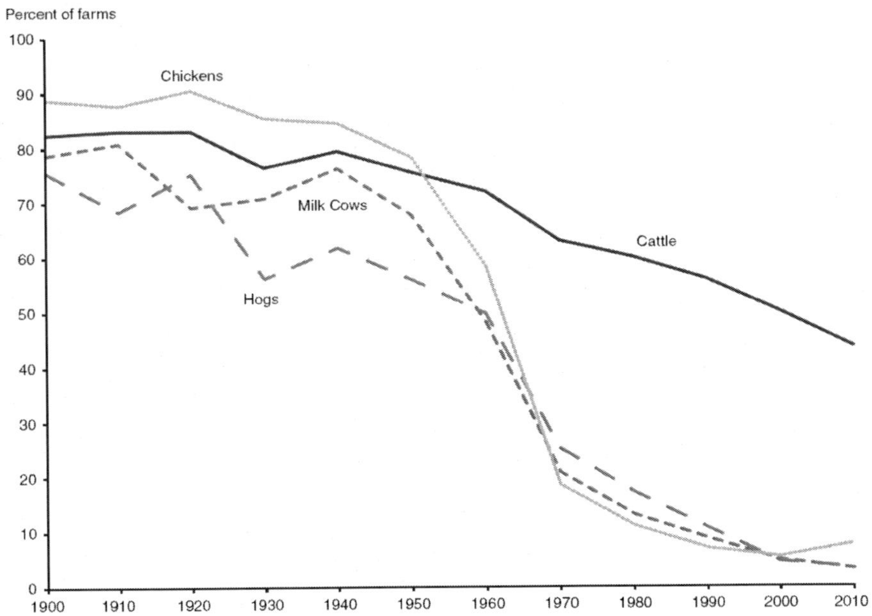

Source: Census of Agriculture, with data interpolated between Census years.

Figure 6. Livestock on U.S. farms, 1900-2010.

However, specialization carries risks. Specialization may create greater production risks from pest outbreaks, as well as economic risks from concentrated price shocks. Diversification across multiple crops might allow the farmer to reduce the financial risks of relying on one commodity, if commodity returns are not highly correlated across crops. Diversification can provide production advantages if farmers can reduce pest presence and improve soil attributes by rotating different crops through field. Finally, diversification may also allow farmers to more fully utilize their labor and capital over the course of a year, and thereby reduce production costs, if different crops are planted and harvested at different times of the year.

U.S. crop farms tend to resolve the trade-off between specialization and diversification by focusing on the production of a few crops—few farms are completely specialized, but few are highly diversified. Table 13 summarizes product diversification on crop farms in 2011. It reports data for all crop production, and for production of 18 different crop categories.[23]

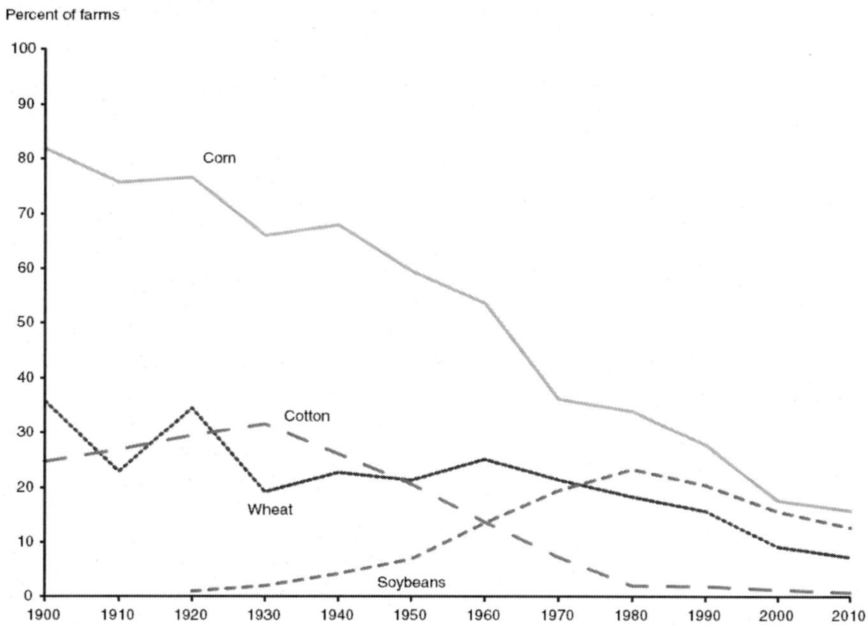

Source: Census of Agriculture, with data interpolated between Census years.

Figure 7. Selected field crops on U.S. farms, 1900-2010.

Twenty-seven percent of crop production occurred on farms that also had livestock—that is, 73 percent occurred on farms that specialized only in crops (table 13). The incidence varied across commodities—for example, barley, hay, oats, and sorghum producers frequently raise livestock (usually cattle), while little fruit and nut, greenhouse, sugar, rice, or vegetable production occurred on farms with livestock.

Table 13. Diversification in U.S. crop production, 2011

	Percent of 2011 value of crop production originating on farms with:					
Commodity	1 crop	2 crops	3 crops	4 crops	More than 4 crops	Livestock
All crops	22	30	22	15	11	27
Barley	4	19	33	17	27	43
Canola	0	11	25	20	44	23
Corn	4	42	25	15	14	34
Cotton	12	18	22	25	23	18

Table 13. (Continued)

Commodity	Percent of 2011 value of crop production originating on farms with:					
	1 crop	2 crops	3 crops	4 crops	More than 4 crops	Livestock
Fruits/nuts	84	8	2	1	5	7
Hay	33	23	16	13	15	42
Nursery/greenhouse	87	8	3	1	1	3
Oats	1	3	24	39	33	55
Peanuts	3	16	30	22	29	20
Potatoes	17	13	20	16	34	19
Rice	30	28	23	11	8	6
Sorghum	1	19	28	24	28	51
Soybeans	3	42	23	17	15	33
Sugar beets	0	2	16	40	42	25
Sugar cane	83	7	8	2	0	8
Tobacco	30	10	18	25	17	40
Vegetables/melons	29	15	16	16	24	9
Wheat	15	19	27	20	19	35

Note: Crops are those listed in section B of the ARMS Phase III questionnaire: barley; canola; corn (grain and silage combined); cotton; dry edible beans/peas/lentils; fruits, nuts, and berries; hay (alfalfa and all other combined); nursery and greenhouse crops; oats; peanuts; potatoes; rice; sorghum (grain and silage combined); soybeans; sugar beets; sugarcane; tobacco; vegetables (processing and all others combined); wheat; other oilseeds; and all other crops.

Source: 2011 Agricultural Resource Management Survey, Phase III, all versions.

Twenty-two percent of crop production occurred on farms that produced only a single crop commodity, while 30 percent occurred on farms with two crops. Only 11 percent occurred on farms with five or more crops. Note that the analysis uses just 21 crop categories, and that some categories—such as vegetables or fruits—contain many specific crops. Table 13 only captures farm diversification across categories, so production of multiple vegetable crops is not captured in this measure.

Farms that produce major field crops rarely specialize in just one—note that less than 5 percent of corn and soybean production occurred on farms with a single crop (that is, only corn or only soybeans). Two-fifths of production was on farms with two crops (typically corn and soybeans, in rotations designed to maintain soil quality and limit pest infestations), while another quarter occurred on farms with three crops. Moreover, over a third of corn and soybean production was on farms that also had livestock production; on many of these farms, livestock manure is "fed" as fertilizer to crops, while the crops are sold and feed is purchased.

Substantial production of hay (33 percent) and tobacco (30 percent) occurs on farms that grow only those crops, but there is also a considerable amount of production on farms with four or more crops. Sugar cane, fruits and nuts, and nursery/greenhouse operations are highly specialized, but farms can and often do grow multiple fruit or nursery crops.

Contracts and Crop Production

Greater specialization can expose farms to greater financial risks. Today, many farm product transactions are organized through contracts between farmers and buyers that are reached prior to harvest (or before the completion of a production stage, as in the case of livestock), and that govern the terms under which products are transferred from the farm. Contracts can reduce financial risks for farmers by providing a more secure outlet for production and by reducing the farmer's exposure to input and output price fluctuations. Consequently, contracts may also expand the supply of credit available to producers by providing lenders with the assurance of outlets for farm products, thus reducing the risks of lending (Key, 2004). In each case, the expanded use of agricultural contracts may facilitate shifts to larger farm operations.

In ARMS, ERS distinguishes two types of agricultural contracts: production and marketing contracts. *Production contracts*, used primarily in livestock production but also common in vegetables and horticulture, specify services to be provided by a farmer for a contractor who owns the commodity while it is being produced. Contracts cover (1) specific services to be provided, (2) the manner in which the farmer is to be compensated for the services, and (3) specific contractor responsibilities for provision of inputs.

Marketing contracts are far more important in crop production. They focus on the commodity as it is delivered to the contractor, rather than specify the services to be provided by the farmer. They set a commodity's price or a mechanism for determining the price, a delivery outlet, and a quantity to be delivered. Forward contracts, with a specific price set at the time of agreement, are one type of marketing contract. But other types, often of longer duration, specify a method or formula for determining prices, rather than a specific price, at the time of agreement. The pricing mechanisms may limit a farmer's exposure to the market price risks. They usually specify minimum acceptable levels of various product attributes, and they often specify price premiums to be paid for desired levels of attributes (such as oil content in corn), thereby providing incentives to produce higher cost but desirable product varieties.

Contracts covered 40 percent of all agricultural production in 2011 and 32 percent of crop production, where their use is growing; contracts covered 28 percent of crop production in 2001-02 and 23 percent in the mid-1990s (MacDonald and Korb, 2011). Earlier data on contract use in agriculture is sparse, but the Census of Agriculture reports that contracts covered 11 percent of all agricultural production in 1969, which suggests that there was an important shift toward greater reliance on contracts since then.

Does contract use affect farm size and structure? Larger crop farms are more likely to use marketing contracts (see also Key, 2004). Table 14 compares attributes of farms that use marketing contracts to those that do not, in each of eight major crop categories. Contracting farms tend to be larger: they operate substantially more farmland than farms that do not contract; they have higher values of production; and they harvest more acres of each crop. Corn, soybean, and wheat farmers who contract tend to place a little over half of their crop production under contract, often in multiple contracts with different buyers. They also combine contracting with cash sales, sales from storage, on-farm feeding, and financial hedging in a broad risk-management strategy. Other contract producers commit almost all of their crop production to the contract, usually with a single buyer. In each commodity, farmers who use contracts for one crop also contract extensively for their other crops, while noncontract producers rely on cash markets. For large-scale producers, contracts form part of a portfolio of marketing strategies to manage risk.

Table 14. Contract and noncontract U.S. field crop producers, by commodity, 2011

Crop and contract status	Whole farm attributes		Reference crop acres	Share under contract	
	Farmland acres	Value of production ($)		Reference crop	Other crops
Corn			*Mean values*		
Contract	911	690,148	409	53.6	40.7
Noncontract	545	338,795	208	0.0	4.7
Cotton					
Contract	1,593	1,271,321	669	81.6	56.9
Noncontract	1,276	812,944	413	0.0	16.3
Fruits/nuts					

Crop and contract status	Whole farm attributes		Reference crop acres	Share under contract	
	Farmland acres	Value of production ($)		Reference crop	Other crops
Contract	170	619,268	109	96.2	31.3
Noncontract	91	217,220	41	0.0	2.8
Peanuts					
Contract	1,509	1,037,972	240	99.0	55.8
Noncontract	665	544,094	127	0.0	15.1
Rice					
Contract	1,587	1,234,746	593	85.7	52.6
Noncontract	1,197	767,206	413	0.0	29.9
Soybeans					
Contract	920	652,975	380	56.3	43.1
Noncontract	513	320,672	196	0.0	5.3
Vegetables/melons					
Contract	1,386	1,631,856	136	96.2	34.0
Noncontract	126	209,557	24	0.0	0.0
Wheat					
Contract	468	872,068	468	59.7	45.9
Noncontract	271	484,596	271	0.0	11.0

Note: The estimates are drawn from farms with any harvested acreage in the listed crops, with 5,109 observations for corn, 693 for cotton, 1,445 for fruits and nuts, 204 for peanuts, 284 for rice, 5,107 for soybeans, and 3,590 for wheat. The estimate for the share of other crops under contract is based on farms that produced other crops in addition to the reference crop.

Source: 2011 Agricultural Resource Management Survey, Phase III, all versions.

The Location of Crop Production

As livestock and feed production separated, they became concentrated in different regions, and changing regional specialization played a role in changes in farm size. States in the Northeast, Appalachian, and Southeast regions held 18.5 percent of all cropland used for crops in 1950, and 11.2 percent in 2007, a 7.3 percentage point decline that in part reflected declines in cotton and tobacco acreage (table 15). Corn Belt and Northern Plains States increased their share of U.S. cropland from 43 percent to 50 percent.

These long-term acreage adjustments reflected the organizational shifts noted earlier, and also reflect important regional shifts in commodity orientation: as farms became more specialized, Corn Belt States concentrated more heavily on the production and sale of feed crops, and livestock production moved from the Corn Belt to Mountain, Southeastern, and Southern Plains (Hart, 2003). Two regional belts of States illustrate the shifts (table 16).

In 1950, crops accounted for less than a third of cash receipts from farming in the four Corn Belt States of Illinois, Indiana, Iowa, and Ohio. Most cash receipts came from sales of livestock fed on the crops grown on those farms. Corn and soybeans accounted for 63 percent of receipts from crop sales. Sixty years later, livestock sales were no longer dominant: crops accounted for 68 percent of total cash receipts, and corn and soybeans for 93 percent of crop receipts. Those four States accounted for a quarter of all U.S. cash receipts from U.S. livestock production in 1950, but that share fell by half over the next 60 years as beef and dairy production moved out of the region. Those changes reflected a movement of acreage into corn and soybeans, but also an expansion of hog and cattle production in other regions and the rapid growth of U.S. poultry production, carried out largely in the South.

Table 15. Regional shifts in cropland, 1950-2007

Item	1950	1982	2007
Total acres (millions)	383.2	382.8	334.9
	Percent of U.S. cropland used for crops		
Region:			
Northeast	5.5	3.6	3.3
Appalachian	6.8	5.0	5.1
Southeast	6.2	3.8	2.8
Delta	4.8	5.0	4.5
Lake States	10.6	10.4	10.8
Corn Belt	21.0	22.6	24.7
Northern Plains	22.2	24.5	25.2
Southern Plains	11.4	9.6	9.2
Mountain	7.0	9.8	9.2
Pacific	4.5	5.7	5.2
All	100.0	100.0	100.0

Note: "Total acres" refers to cropland used for crops (and harvested, failed, or left fallow), and excludes cropland used only for pasture or left idle.
Source: ERS Major Lands Uses, http://www.ers.usda.gov/Data/MajorLandUses/

A near mirror shift occurred in a belt of six Southern States of Arkansas, Mississippi, Alabama, Georgia, South Carolina, and North Carolina (table 16). Crops accounted for 72 percent of cash receipts in 1950, and most of that reflected just two crops—cotton and tobacco. By 2010, crops fell to 40 percent of cash receipts as cotton and tobacco declined and poultry and hog production expanded. Together, those States accounted for 17 percent of all U.S. livestock receipts in 2010, compared to 5 percent in 1950. While feed crops are grown in those States, they import much of their feed from the Corn Belt.

Table 16. Shifts in agricultural specialization in selected States, 1950-2010

Item	1950	1980	2010
		Percent	
Crop share of region cash receipts Corn Belt 4	29	61	68
Corn and soybean share of crop receipts	63	89	93
Southern 6	72	51	40
Cotton and tobacco share of crop receipts	71	32	18
Share of U.S. livestock cash receipts			
Corn Belt 4	26	16	13
Southern 6	5	10	17
Share of U.S. crop cash receipts			
Corn Belt 4	13	23	22
Southern 6	18	10	9

Note: "Corn Belt 4" is Illinois, Indiana, Iowa, and Ohio, while "Southern 6" is Alabama, Arkansas, Georgia, Mississippi, North Carolina, and South Carolina.

Source: ERS Farm Income database (www.ers.usda.gov/Data/FarmIncome/finfidmu.htm)

The interregional shifts in cropland affected crop farm size. Cropland declined in regions with hilly topography and mixed land use—cropland interspersed with forests, residences, and commercial uses—and shifted toward regions with flatter land and more of the land base devoted to crops. The latter were more suited to the types of equipment that could allow a farmer to manage a much larger farm. As a result, interregional shifts in cropland supported the impacts of new technology in driving production to larger farms.

Drivers of Consolidation: Government Policies

Federal policies may affect farm structure through multiple channels. Some programs may increase consolidation, while others may reduce consolidation. Some effects are straightforward and fairly direct, while others are subtle and indirect.

Tax policies can affect prices for capital goods, thereby affecting purchases of farm machinery and structures. Durst (2009) notes that the treatment of agriculture in the tax code, through capital cost recovery provisions, has effectively reduced the costs of capital to agriculture. LeBlanc and Hrubovcak (1986) found that changes in tax policy had important impacts on the cost of capital and equipment investment in agriculture. If lower tax rates lead to lower costs of capital, and if farmers respond by purchasing more capital equipment, then tax policy can enhance technology's effect on farm size.

USDA lending programs extend credit to farmers who may not be able to obtain loans through commercial channels, and some programs also provide subsidized interest rates. To the extent that these programs either help new small farms become established or prevent existing small farms from closing, they can limit cropland consolidation.

USDA research programs may also affect the availability and prices of technologies, and hence, farm sizes. Schmitz and Seckler (1970) analyze one well-known example, the development of the mechanical tomato harvester. Government provided about 40 percent of the total research and development expenses for the harvester, which once introduced led to the replacement of farm labor with capital and thereby induced a shift to larger operations. Similarly, Fernandez-Cornejo and Pho (2002) argue that public and private research programs to produce improved herbicides generated a labor-saving innovation that allowed farmers to operate larger farms with the hours saved from weed control.

Regulatory policies aimed at environmental or food safety targets may alter the relative costs of large and small farms, thereby inducing changes in farm structure. For example, Sneeringer and Key (2011) showed that, after the Environmental Protection Agency imposed size-based thresholds for regulation of concentrated animal feeding operations (CAFOs) under the Clean Water Act, newly constructed operations tended to enter the hog industry at sizes just under the regulatory threshold. During this period, scale economies in hog production were leading to much larger operations (Key and McBride, 2007), so the CAFO rules induced some farmers to constrain farm size, compared to what they would have developed absent the regulation.

Issues related to farm structure played a role in the design of the Food Safety Modernization Act (FSMA) of 2011, which gives the Food and Drug Administration new enforcement authorities to improve compliance with risk-based food safety standards. The rules apply new standards for growers of fruits, vegetables, nuts, mushrooms, herbs, and sprouts. When the rules are fully implemented, they will impose higher costs on farms not already in compliance. The bill includes new requirements for testing and reporting that would likely have imposed higher costs, per unit of production, on smaller farms. The Act exempts certain farms from compliance—those that sell the majority of their products directly to consumers or retail outlets within 275 miles or within the same State, and that have annual sales of less than $500,000.

Federal commodity programs have been a key focus of discussion among analysts of farm consolidation. The Government operates broad-based commodity programs that affect a range of field crops, as well as several programs aimed at individual commodities. The programs' designs create varied impacts on farm structure.

Specific Commodity Programs: Marketing Quotas

In the case of peanuts and tobacco, marketing quotas regulated prices and production until the early 2000s and appear to have limited farm size (Dohlman et al., 2009; Kirwan et., 2012). Marketing quotas set limits on aggregate supply by giving quota owners the exclusive right to sell a set amount of the commodity at or above a support price within a geographic area. Tobacco producers without quota could not sell the commodity, so tobacco marketing quotas also fixed annual supply, while peanut producers without quota faced marketing restrictions that left them selling at lower prices to the export and crush markets (Dohlman et al., 2009). Because support prices were set well above cash expenses, quotas became valuable assets that could be exchanged.

The transfer of quota rights among quota owners was restricted. In most States, tobacco quotas could not be exchanged across county lines. Peanut restrictions were more complex, but transfers of quotas across State lines were generally forbidden, and transfers across county lines were limited. The programs had been in place with some adjustment since 1930 for tobacco and the 1940s for peanuts. Since then, production of other crops has shifted to much larger farms while also undergoing a series of locational adjustments. Because quotas made such adjustments more difficult and costly, they limited geographic adjustment, growth in farm size, and productivity growth.

The peanut quota system was eliminated in 2002, and the tobacco system was eliminated in 2004. Each policy change featured buyouts of quota owners. Peanut production came under broad-based commodity program support, while all Federal support for tobacco producers was ended.

Each crop saw rapid shifts of production to larger farms and different regions after marketing quotas were eliminated; the shifts to larger farms were much stronger than any previous trends in the sectors, and much stronger than shifts in other commodities. In 2001, just before the buyout, the average peanut farm operated 860 acres in total, with 120 in peanuts; just 6 years later, mean farm size had increased to 1,525 acres, with 227 in peanuts (Dohlman et al., 2009). Average peanut yields per acre rose by 21 percent.[24]

In flue-cured tobacco, mean tobacco acres per farm increased from 33, at the time of the buyout in 2004, to 84 in 2007, while mean farmland acres rose from 566 to 906. Burley tobacco farms are much smaller, but average burley tobacco acreage rose from 5 to 10.5 acres, and mean farm size rose from 191 to 247 acres (Dohlman et al., 2009). In a detailed study of Kentucky tobacco operations, Kirwan et al. (2012), using Census of Agriculture data, found that tobacco farm productivity increased by 44 percent between 2002 and 2007 (a period encompassing the policy change) after declining by 7 percent in 1997-2002. They found that average tobacco farm size doubled during the buyout period, and that structural change played a major role in driving the increase in productivity.

Other factors also contributed to the changes. Many older tobacco producers anticipated the buyout and remained in tobacco farming to be eligible for transition payments made to active growers. As a result, farm closures were lower just before the policy change, and higher just after, than they would have been without transition payments (Dohlman et al., 2009). In peanuts, market prices and planted peanut acreage fell after the buyout, and those shifts had independent effects on peanut acreage per farm. However, in each case, the design of the quota systems suggested that they would have limited farm size, and the rapid shifts of production to much larger farms in the aftermath of quota elimination provide strong evidence that the programs effectively did limit farm size, aside from other more transitory factors.

Broad-Based Commodity Programs

Broad-based USDA commodity programs direct their support primarily to producers of certain field crops: wheat, feed grains, cotton, rice, and oilseeds, and some programs also cover pulse crops, honey, and wool. A redesign of the programs in the mid-1990s made direct payments (DPs) to farmers based upon the acreage they historically enrolled in commodity programs and the historic crop yields from that acreage. The programs also provide countercyclical payments (CCPs), on the same production basis, when market prices fall below target levels and provide for marketing assistance loans, based on current production, that offer support when prices are below target levels. Marketing assistance loans, which effectively set minimum prices that farmers can expect to receive for their crops, have been part of commodity programs in one form or another since the 1930s, while DPs and CCPs were more recent developments.[25]

Broadly speaking, payments tend to follow production: farms with greater production of a program commodity—either because they have more acreage of the commodity or higher yields on their acreage—receive greater total payments. There are some limits on the aggregate payments that farm operators can receive, but the limits have not affected many existing operations (White and Hoppe, 2012). Despite receiving higher overall payments, larger farms do not receive systematically higher payments on a per-acre or per-bushel basis than small farms receive, nor do commodity payments represent a higher share of their gross income. Moreover, payments tend to be capitalized into land values, and to raise cropland rents, thus benefiting landlords and raising renters' costs. They do not provide an obvious direct impetus to increase farm size.

However, several authors have argued that commodity programs contributed indirectly to the explosion of agricultural productivity growth, and the concurrent and ongoing shifts of production to larger farms that followed the New Deal introduction of the programs, because of their effects on farm financial risks, farmer behavior, and lenders. Cochrane and Ryan (1976, p. 373) argued that commodity programs "provided the stable prices, hence price insurance, to induce the alert and aggressive farmers to invest in new and improved technologies and capital items, and the reasonably acceptable farm incomes and asset positions to induce lenders to assume the risk of making farm production loans."

Clarke (1994) offered a specific application of the Cochrane-Ryan hypothesis focused on the diffusion of the tractor in the 1920s and 1930s. She developed a model of the value of adopting tractors on farms, and argued that farms with more than 100 acres of land should have adopted the tractor during the 1920s. Yet many did not, and Clarke argues that the riskiness inherent in farming at the time prevented adoption; before the introduction of New Deal commodity programs, farm prices and incomes varied widely from year to year. In addition, credit markets were focused on short-term loans, and many farmers did not have access to credit—in more modern language, they were "credit-constrained." Farmers, therefore, hesitated to make the long-term financial commitment necessary to purchase durable goods like tractors.

The introduction of New Deal commodity programs, according to Clarke, reduced farmers' financial risks by setting floors below which commodity prices would not fall. New Deal lending programs made credit more available and also gave competition to commercial lenders, who reduced rates and broadened terms in response to the competition. Improved credit and reduced financial risks led farmers to purchase capital goods and also to make cash outlays for intermediate goods like commercial fertilizers, prepared feeds, and hybrid seeds. With increased and more assured demand for capital equipment, manufacturers invested in improvements to equipment.

Clarke (1994) and Cochrane and Ryan (1976) argued that commodity programs complemented and accelerated the spread of labor-saving technology, by hastening the adoption of capital equipment, and may also have affected the types of farmers who adopted new equipment. That position is not undisputed; others see a far greater role for technological advances, often fueled by Government investment in research, in driving productivity growth and farm structure during the New Deal and thereafter.[26] Moreover, farm commodity programs have changed considerably since the period that they analyze. In particular, commodity programs since the 1980s have been built around DPs to farmers, and income stabilization, instead of supply controls aimed at supporting commodity prices (Dimitri et al., 2005).

However, the historical analyses are valuable because they focus on the effects policy has on risks; on the responses of farmers, input providers, and lenders to programs that reduce or manage farm-level risks; and on the consequent effects on farm structure. Modern commodity programs still clearly target the financial risks of farming (Westcott and Price, 2001). With lower risks, farmers may specialize more and may make more commodity-specific equipment investments, which may in turn lead to farm expansion. DPs may also give assurance of lower risks to lenders, who may be more willing to lend to producers who already have a guaranteed revenue source through commodity payments. Changes in perceived risks may influence farm structure, by influencing the adoption of technology as well as farmers' production, investment, and work decisions.

Recently, Key and Roberts (2007a; see also Key and Roberts, 2007b, and Roberts and Key, 2008), hereafter KR, developed an empirical analysis of modern links between farm program payments and consolidation. They used Census of Agriculture data to calculate midpoint acreages for cropland at a highly granular level—postal zip codes—for 1987, 1992, 1997, and 2002.[27] They also used census data to calculate total Government payments per cropland acre, from commodity and conservation programs, for each zip code.

Payments per acre vary widely across zip codes, depending on the composition of crops grown in the area as well as historic yields from those crops.

KR found that cropland consolidated more rapidly in those zip codes with higher levels of Government payments per acre at the beginning of the period. For example, during 1987-2002, the midpoint acreage farm size increased by 46.3 percent in zip codes with the highest levels of payments, compared to 23.6 percent in zip codes with the lowest levels of payments and 11.2 percent in zip codes with no payments.[28] The association between payments per acre and the growth in midpoint acreages was statistically significant and large. It was also robust: growth rates increased consistently from one payment class to the next, and the relationship held in each time period (1987- 1992, 1992-1997, and 1997-2002, as well as 1987-2002).

As part of the same research project, KR (2007b), analyzed the linkage between payments and farm business survival. They focused on wheat, corn, soybean, and cash grain (diversified) crop farms and found that farms with higher 1987 levels of Government payments per acre were more likely to still be in business in subsequent censuses. Surviving farms were also likely to grow faster the higher the 1987 levels of per-acre Government payments they received. Each effect was stronger for farms that started out larger, with size and growth measured in acres of farmland.[29]

KR identified a robust association between Government payments and later consolidation, but did higher Government payments cause faster consolidation? Three factors weigh against a causal interpretation:

Conflicting evidence at the commodity level. Midpoint acreage for fruit, vegetable, and livestock commodities that are not covered by commodity programs increased during 1987-2007 at rates comparable to those for field crops (tables 2-4). That pattern suggests that factors other than payments must play an important role.

Marginal association between payments and consolidation. Much of the cropland in zip codes with low levels of payments per acre is idle, devoted to pasture and grazing, or in land retirement programs. These zip codes show very little consolidation. However, in the zip codes where most planted cropland is (and most consolidation), once payments per acre reach a minimal level, differences in payment rates have a much weaker association with consolidation.

Omission of alternative drivers. The highest rates of consolidation occurred in the Plains, Corn Belt, and Delta. They have the highest values of Government payments per acre because they have heavy concentrations of program crops and high yields on crops. They also have large, flat, and contiguous fields--land best suited for the labor-saving innovations described earlier. Although KR included several control measures (sales per cropland acre, the fraction of land devoted to cropland, beginning-of-period median size, and locational controls), there is still reason to believe that the association between payments and growth may reflect the greater adoption and use of labor-saving innovations in those regions.

KR focused on the association between Government payments and later cropland consolidation between 1987 and 2002. Significant changes in policy and in commodity markets have occurred since then. In particular, as a result of higher commodity prices and Federal spending limits on commodity programs, commodity programs payments now account for a much smaller share of the value of crop production (fig. 8). Between 1999 and 2001, Federal commodity program payments amounted to over 20 percent of the value of U.S. crop production. If the value of fruit and vegetable production is subtracted from crop production (since those commodities do not receive support), then commodity payments came to over 30 percent of the value of production. After 2005, commodity payments fell sharply, as a share of production, and by 2011, they amounted to just over 3 percent of the value of crop production (or 4 percent of crop production, excluding fruits and vegetables). Over the same period, crop insurance subsidies rose sharply and by 2011 exceeded commodity program payments.

Crop Insurance

The Federal crop insurance program has become a major component of agriculture support via premium subsidies and support for administrative costs provided by the Federal Government. Federal subsidies for premiums totaled $7.15 billion in 2012, and Federal support for insurance company expenses were $1.38 billion, by which time the program covered 282 million acres (Shields, 2012).

Federal crop insurance was created in 1938 but was limited in scope until Congress expanded the program in 1980 to cover many more crops and regions; in 1994 and 2000, Congress expanded premium support through subsidies and a wider range of commodity coverage. By 2012, over 80 percent of planted acreage of corn, cotton, soybeans, and wheat were covered by crop insurance (Shields, 2012; Glauber, 2013).

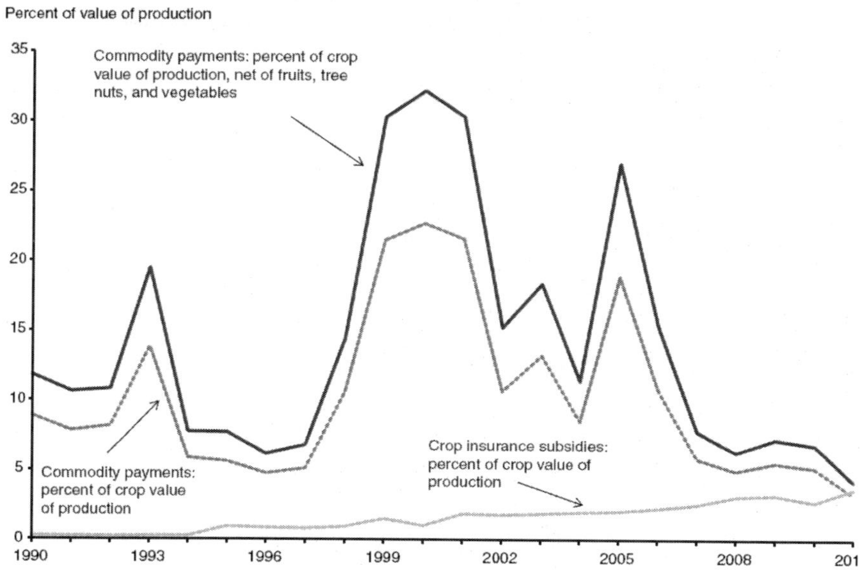

Sources: USDA Economic Research Service for direct payments and value of crop production; Glauber (2013) for Federal crop insurance subsidies.

Figure 8. Commodity program payments and crop insurance subsidies, relative to the value of crop production.

The Federal Crop Insurance Reform Act of 1994 (FCIRA) led crop insurance coverage to increase in the mid-1990s, from about 100 million acres of cropland in 1994 to 220 million the next year. Enrollment fell to 180 million acres in 1997 and 1998 as some initial mandatory enrollment provisions of the 1994 legislation were dropped, but the legislation nonetheless had a major impact on enrollment, which expanded steadily from 1998 to 2008.

The FCIRA provided a natural experiment for analysts interested in farmer responses to risk: expanded premium subsidies led to suddenly reduced production risks for farmers, which varied for different crops and different locations. In two papers, O'Donoghue, Key, and Roberts evaluated how the change affected farmer behavior (O'Donoghue et al., 2009; Key et al. 2006). Farmers can manage the income risks associated with farming by diversifying their sources of income. Working off the farm leaves a farmer's household income less exposed to fluctuations in farm performance. Likewise, producing a range of commodities leaves household income less exposed to fluctuations in the performance of a single crop. Each strategy carries costs of its own.

Expanded crop insurance, by reducing the financial risks of farming, may lead farmers to reduce their other risk management activities.

In O'Donoghue et al. (2009), farms with expanded insurance coverage reduced their levels of commodity diversification slightly—that is, they specialized more—as greater insurance subsidies became available. In Key et al. (2006), large farms reduced their hours of off-farm work after the 1994 legislation, while small farms increased off-farm work. Because off-farm work is a substitute for onfarm work, one can infer that larger farm operators increased their hours devoted to farm work, while smaller operators reduced their farm hours.[30] With more hours to apply to farming, a farmer will likely expand the farm. The net effect would have been an increase in farm consolidation. While the effect on commodity diversification was small, the effect on labor was more substantial—up to 22 days per year for larger farms. These studies likely captured a short-run response, since they covered changes between 1992 and 1997, with the new legislation being passed in 1994. Major changes in farm size—which requires assembling land, capital equipment, and labor—may take longer.

A third study (Claassen et al., 2011) modeled the combined impact of Federal disaster programs, crop insurance, and commodity program support on land use changes, through farmer decisions to convert grasslands to cultivated croplands in North and South Dakota. They found that the programs raised the returns to cultivation, thereby increasing cultivated cropland by 2.9 percent between 1998 and 2007. The largest effect was from disaster assistance (1.2 percent) followed by crop insurance (1.0 percent) and marketing loan benefits from commodity programs (0.7 percent). By shifting more land from grassland to cropland, the programs would also increase midpoint cropland acreage by a similar modest amount.

Agricultural production is subject to substantial production and price risks. To manage the risks associated with greater specialization and substantial investments in capital equipment, many farmers look to institutions for support. Some strategies to manage farm risks, like contracting, have been developed in the private sector. Government commodity and crop insurance programs give additional tools for managing farm financial risks. To the extent that the programs reduce risks facing farmers, they may encourage farmers to invest more in labor-saving innovations and to specialize more in agriculture and in specific commodities, which in turn could affect farm structure and help shift production and acreage to larger farms.

THE FUTURE: WILL FAMILY FARMS STILL DOMINATE CROP PRODUCTION?

The largest U.S. farms are now very large indeed. For example, a large cattle feedlot firm operates 10 feedlots, with a combined capacity of 520,000 head; an Oregon dairy farm has a milking herd of 16,000 head, and also farms over 30,000 acres of cropland for forage, wheat, and vegetables; a Florida firm farms over 180,000 acres of sugar cane; and a California firm farms 150,000 acres, primarily in cotton and tomatoes, in that State and another 30,000 acres in Australia.[31] Each has hundreds of employees and teams of professional managers; none are owned and operated by a single family.

These are exceptional operations, easily identified through Internet searches. But there are also many less well-known but very large farming operations: 1,140 U.S. farms had at least 10,000 acres of cropland in 2011, up from 409 in 2001 (each estimate is based on ARMS data). Do the largest farms simply represent the size of a farm that a family can now manage, and that many more families will manage in the future, or do they signal a more fundamental shift toward more corporate, complex, and bureaucratically organized farm firms?

This is not a new question. Fitzgerald (2003) summarizes the enthusiasm for large, bureaucratically organized farming corporations expressed by some economists, engineers, businessmen, and political leaders during the 1920s. Those firms often operated multiple farms with hired farm operators and skilled professionals on staff; invested significant amounts in machinery, structures, and transportation links; and developed detailed operating protocols and production plans to be applied across their farms.

They were organized much like the "bonanza" wheat farms in the Red River Valley in Minnesota and North Dakota in the late 19th century (Allen and Lueck, 2002). Bonanza farms were integrated into all stages of production, from sod-busting to processing, and were highly mechanized, with the latest large-scale farm machinery. Owned by absentee investors, most were managed in multiple units, with divisions of approximately 5,000 acres as well as subdivisions of 1,200 to 1,800 acres. Managers received salaries and profit-sharing commissions, manual laborers were paid a daily wage and worked under rigid rules, and technical specialists at the firms' headquarters received salaries.

Between 1875 and 1890, bonanza farms were generally profitable and were hailed as the future of agriculture. Eventually, however, the farms became unprofitable, and most had disappeared by the 1920s. Similarly, few of the farms touted by the subjects of Fitzgerald's book survived the Great Depression. Family farms accounted for most agricultural production in the period, and they continue to do so in the United States today despite the major changes in farm structure and despite the differing organizational structures used by the very largest farm firms.

Family Farms in Today's Crop Agriculture

ERS defines a family farm to be one in which the principal operator, and people related to the principal operator by blood or marriage, own more than half of the farm business.[32] Family farms are defined by ERS only by ownership and operation, and not by size or by labor commitments. Large family farms rely heavily on hired labor, rented land, and contracted services to operate their businesses. However, the definition makes an important point: businesses owned and operated by family groups—a specific and distinctive type of organization—continue to dominate agricultural production in the United States, even as production has shifted to much larger family businesses.

Table 17. Family farms in crop production, 2011

Commodity	Family farm share	
	U.S. farms	Value of production
	Percent	
All crop production	96.4	86.6
Corn	97.2	94.2
Cotton	96.0	91.5
Peanuts	93.8	84.0
Rice	89.4	88.2
Soybeans	97.2	94.8
Wheat	96.7	93.5
Vegetables/melons	96.0	69.3
Fruits/nuts	92.8	65.7
Nursery	97.0	70.3

Note: Family farms are defined as those in which the principal operator, and people related to the principal operator by blood or marriage, own more than 50 percent of the farm business.

Source: 2011 Agricultural Resource Management Survey, Phase III, all versions.

Today, 96 percent of U.S. farms with crop production are family farms, and they originate 87 percent of the value of crop production (table 17). Non-family farms account for less than 6 percent of corn and soybean production; they are far more important to fruit, vegetable, and nursery production, where the 3-7 percent of farms that are non-family operations account for 30-35 percent of the value of production (implying that some are very large farms). But even there, family farms handle 65-70 percent of the value of production.

Family farms can be sole proprietorships, but they are also often organized as partnerships or corporations (table 18). Non-family farms include those operated by cooperatives, by hired managers on behalf of non-operator owners, by large corporations with diverse ownership, and by small groups of unrelated people (often in partnerships or corporations).

There is no evidence of any systematic decline in family operations: the shares of farms and value of crop production held by family farms show no trends since the initial ARMS in 1996. Although none of the four large farm firms noted at the beginning of this section above could be classified as a family farm, 86 percent of farms with at least 10,000 acres of cropland are family operations.

While there are large and complex nonfamily operations in U.S. crop agriculture, and some have been organized recently, there are also examples of large nonfamily corporations that have left farm production in recent years and moved to a coordinating role. For example, Dole Foods once operated extensive farming operations in California and Arizona. Today, it leases 14,000 acres in those States from landowners who purchased the land from Dole. Most of that land is now farmed by independent growers, most of which are family operations, under contract arrangements with Dole. Fresh Del Monte, a major competitor of Dole in fresh fruits and vegetables, has taken a similar approach on U.S. land that it once farmed and now sources primarily from contract growers.[33] These more complex arrangements are shifting agricultural production to family operations, while leaving large distributors as coordinators of production and providers of inputs.

The success of family farms in agriculture arises from several factors. First, although technological scale economies exist in many farm production processes, they are rarely large enough or extensive enough to be beyond the scope of a family business (Deininger and Byerlee, 2012). Even in industrialized livestock sectors, where large corporations coordinate most hog and broiler production through their role as integrators, most production is still carried out on family operations under contract with the integrators. Measured scale economies in farm production, while important, appear to be exhausted or largely realized at still-modest sizes.[34]

Table 18. Family and nonfamily crop farms, by legal status, 2011

	Family farms		Nonfamily farms	
Legal status	Farms	Production	Farms	Production
	Percent of all family or nonfamily farms			
Sole proprietorship	92.1	66.5	46.5	11.1
Partnership	3.8	16.7	24.8	31.3
Corporation	3.1	15.4	17.7	49.6
Other	1.0	1.4	11.0	8.0
	100.0	100.0	100.0	100.0

Note: Production refers to the value of crop production. The legal status "other" includes estates, trusts, cooperatives, grazing associations, and public agencies.
Source: 2011 Agricultural Resource Management Survey, Phase III, all versions.

Second, analysts widely believe that managerial diseconomies of scale set in as farms get large enough to have to rely extensively on hired managers and workers: the firm's owners must then closely monitor the decisions made by hired staff (Allen and Lueck, 2002; Deininger and Byerlee, 2012; Gardner, 2002). These diseconomies are thought to be particularly acute in agriculture, because of the degree of local knowledge of soils, nutrients, pests, and weather patterns required to effectively manage cropping and animal husbandry decisions, and because sudden changes in weather or in animal performance demand quick and informed reactions.

Challenges for Family Farms

Despite the historical predominance of family farms as an organizational mode, several factors could undermine this tendency. First, even though families can manage much larger operations than in the past, the land and capital equipment requirements for these businesses may create unacceptable financial risks for many families. With Corn Belt cropland selling for $7,000 an acre in 2012, a farmer aiming to produce 600 acres of corn and 500 of soybeans, close to the midpoint acreages of table 2, would need to assemble over $8 million of land, equipment, and structures. Irrigated California cropland was valued at $12,000 per acre in 2012; a farmer with 300 acres of irrigated fruits and vegetables would need to assemble over $4 million in land, equipment, and structures.[35] Much of this capital may be leased or rented, but the equity and debt required for the owned assets may be daunting; even if feasible, most or all of a family's wealth could be tied up in one risky enterprise.

The risks have become more salient in recent years, with sharp fluctuations in global commodity prices for crops and for energy inputs (Baffes, 2013). While public policies give some insulation in the form of commodity and crop insurance programs, and private institutions in the form of contracting for inputs and products, these movements represent unprecedented volatility for many producers. As a result, operators of large family farms may show more interest in different ways of organizing farm production and accessing capital. They are likely to explore more complicated ownership and management structures for large family farms, with non-operator equity and resource providers, the sharing of assets and services among farms, and more complex contractual relationships with input suppliers and with processors.

A second deterrent to would-be family farms may be created by new precision technologies that erode the managerial diseconomies associated with using hired managers and labor, in turn reducing the advantages of close-knit family operations. In livestock production, specialized confined feeding operations rely on housing with automated climate controls and sanitation equipment, strict biosecurity procedures, specifically formulated purchased feeds, and tightly controlled animal genetics. Those features may have reduced the value of localized farm and herd-specific knowledge and improved the capability for monitoring the performance of hired managers, allowing for greater reliance on hired managers, workers, and nonfamily organization (Allen and Lueck, 2002: Deininger and Byerlee, 2012).

Some see such developments emerging in crop agriculture (Deininger and Byerlee, 2012; Byerlee et al., 2012; Boehlje and Gray, 2009). Large firms now routinely manage businesses with well over 100,000 acres in South America, Ukraine, and Russia. These firms tend to use genetically engineered seeds and no-till cultivation techniques to limit the amount of management time expended on field-level weed, pest, and nutrient management decisions. They also use the communications, measurement, and monitoring capabilities now incorporated in precision agriculture technologies to provide the detailed and localized field and farm level information that was previously available only through persistent personal experience in fields. They have introduced routines, process controls, and standardized operating procedures into field-level farm management, and they employ highly trained scientific and business staff to handle technical tasks. Their size and geographically dispersed operations provide them with opportunities to reduce costs through more intensive use of capital equipment, and they have been able to tap independent investors for financing.

Such organizations remain rare in the United States, where family farms still dominate U.S. crop agriculture even as it has shifted to much larger farms. They will continue to do so as long as they are able to limit and manage the financial risks associated with managing large and capital-intensive businesses, and as long as the strengths of family organizations—localized knowledge, quick and flexible adjustments to changed circumstances, and the incentives to act on each of those—remain necessary to crop production.

REFERENCES

Allen, Douglas W., and Dean Lueck. 2002. The Nature of the Farm: Contracts, Risk, and Organization in Agriculture. Cambridge, MA: The MIT Press.

Baffes, John. 2013. Commodity Market Outlook. Development Prospects Group. The World Bank. January.

Bechdol, Elizabeth, Allen Gray, and Brent Gloy. 2010. "Forces Affecting Change in Crop Production Agriculture." Choices: The Magazine of Food, Farm, and Resource Issues (25): 11-16.

Boehlje, Michael T., and Allen Gray. 2009. "Biological Manufacturing and Industrialization of Production Agriculture." Purdue University. Department of Agricultural Economics Working Paper No. 0912.

Byerlee, Derek, Alex Lissitsa, and Paula Savanti. 2012. "Corporate Models of Broadacre Crop Farming: International Experience from Argentina and Ukraine." Farm Policy Journal (9): 13-25.

Calvin, Linda, and Philip Martin. 2010. The U.S. Produce Industry and Labor: Facing the Future in a Global Economy. U.S. Department of Agriculture, Economic Research Service, ERR -106. November.

Clarke, Sally H. 1994. Regulation and the Revolution in United States Farm Productivity. Cambridge, UK: Cambridge University Press.

Claassen, Roger, Fernando Carriazo, Joseph C. Cooper, Daniel Hellerstein, and Kohei Ueda. 2011. Grassland to Cropland Conversion in the Northern Plains: The Role of Crop Insurance, Commodity, and Disaster Programs. U.S. Department of Agriculture, Economic Research Service, ERR-120. June.

Cochrane, Willard W. 1993. The Development of American Agriculture: A Historical Analysis. 2nd edition. Minneapolis: University of Minnesota Press.

Cochrane, Willard W., and Mary E. Ryan. 1976. American Farm Policy, 1948-1973. Minneapolis: University of Minnesota Press.

Davis, Steven, and John Haltiwanger. 1991. "Gross Job Creation and Destruction: Microeconomic Evidence and Macroeconomic Implications." Microeconomic Brookings Papers on Economic Activity: 115-200.

Deininger, K., and D. Byerlee. 2012. "The Rise of Large Farms in Land-Abundant Countries: Do They Have a Future?" World Development (40): 701-714.

Dimitri, Carolyn, Anne Effland, and Neilson Conklin. 2005. The Twentieth Century Transformation of U.S. Agriculture and Farm Policy. U.S. Department of Agriculture, Economic Research Service, EIB-3. June.

Dohlman, Erik, Linda Foreman, and Michelle Da Pra. 2009. The Post-Buyout Experience: Peanut and Tobacco Sectors Adapt to Policy Reform. U.S. Department of Agriculture, Economic Research Service, EIB-60. November.

Durst, Ron. 2009. Federal Farm Policies and Farm Households. U.S. Department of Agriculture. Economic Research Service. EIB-54. May.

Fernandez-Cornejo, Jorge, and Yvan Pho. 2002. "Induced Innovation and the Economics of Herbicide Use." In Darwin C. Hall and L. Joe Moffitt, eds., Economics of Pesticides, Sustainable Food Production, and Organic Food Markets. Oxford, UK: Elsevier Science.

Fernandez-Cornejo, Jorge. 2007. Off-farm Income, Technology Adoption, and Farm Economic Performance. U.S. Department of Agriculture, Economic Research Service, ERR-36. February.

Fernandez-Cornejo, Jorge, Richard Nehring, Craig Osteen, Seth Wechsler, Andrew Martin, Alex Vialou, and Arthur Grube. 2013. Pesticide Use in U.S. Agriculture, 1960-2008. U.S. Department of Agriculture, Economic Research Service, Forthcoming Economic Information Bulletin.

Fitzgerald, Deborah. 2003. Every Farm a Factory: the Industrial Ideal in American Agriculture. New Haven: Yale University Press.

Florence, P. Sargent. 1933. The Logic of Industrial Organization. London: Kegan, Paul, Trench & Trubner.

Gardner, Bruce L. 2002. U.S. Agriculture in the 20th Century: How It Flourished and What It Cost. Cambridge, MA: Harvard University Press.

Gardner, Justin G., Richard F. Nehring, and Carl H. Nelson. 2009. "Genetically Modified Crops and Household Labor Savings in US Crop Production." Agbioforum 12 (3&4): 303-312.

Glauber, Joseph W. 2013. "The Growth of the Federal Crop Insurance Program, 1990-2011." American Journal of Agricultural Economics (95): 482-488.

Hart, John Fraser. 2003. The Changing Scale of American Agriculture. Charlottesville: University of Virginia Press.

Holmes, Thomas J., and Songhoon Lee. 2012. "Economies of Density vs. Natural Advantage: Crop Choice on the Back Forty." Review of Economics and Statistics (94): 1-19.

Hoppe, Robert, James M. MacDonald, and Penni Korb. 2010. Small Farms in the United States: Persistence Under Pressure. U.S. Department of Agriculture, Economic Research Service, EIB-63. February.

Horowitz, John, Robert Ebel, and Kohei Ueda. 2010. "No-Till" Farming Is a Growing Practice. U.S. Department of Agriculture, Economic Research Service, EIB-70. November.

Key, Nigel. 2004. "Agricultural Contracting and the Scale of Production." Agricultural and Resource Economics Review (33): 255-271.

Key, Nigel, and William McBride. 2007. The Changing Economics of U.S. Hog Production. U.S. Department of Agriculture, Economic Research Service, ERR-52. December.

Key, Nigel, and Michael J. Roberts. 2007a. Commodity Payments, Farm Business Survival, and Farm Size Growth. U.S. Department of Agriculture, Economic Research Service, ERR-51. November.

Key, Nigel, and Michael J. Roberts. 2007b. "Do Government Payments Influence Farm Size and Survival." Journal of Agricultural and Resource Economics (32): 330-349.

Key, Nigel, Michael J. Roberts, and Erik O'Donoghue. 2006. "Risk and Farm Operator Labor Supply." Applied Economics 38: 573-586.

Kirwan, Barret E., Shinsuke Uchida, and T. Kirk White. 2012. "Aggregate and Farm-Level Productivity Growth in Tobacco: Before and After the Quota Buyout." American Journal of Agricultural Economics (94): 838-853.

Kislev, Yoav, and Willis Peterson. 1982. "Prices, Technology, and Farm Size." Journal of Political Economy (90): 578-95.

Larson, James A., and Leslie A. Meyer. 1996. "Supply, Demand, and Prices," in The Cotton Industry in the United States. U.S. Department of Agriculture, Economic Research Service, AER-739. July.

LeBlanc, Michael, and James Hrubovcak. 1986. "The Effects of Tax Policy on Aggregate Agricultural Investment." American Journal of Agricultural Economics (68): 767-777.

Low, Sarah, and Steven Vogel. 2011. Direct and Intermediated Marketing of Local Foods in the United States. U.S. Department of Agriculture, Economic Research Service, ERR-128. November.

Lund, Philip, and Roger Price. 1998. "The Measurement of Average Farm Size." Journal of Agricultural Economics (49): 100-110.

MacDonald, James M., and Penni Korb, 2011. Agricultural Contracting Update: Contracts in 2008. U.S. Department of Agriculture, Economic Research Service, EIB-72. February.

MacDonald, James M., and William D. McBride. 2009. The Transformation of U.S. Livestock Agriculture: Scale, Efficiency and Risks. U.S. Department of Agriculture, Economic Research Service, EIB-43. January.

MacDonald, James M., and Sun Ling Wang. 2011 "Foregoing Sub-therapeutic Antibiotics: the Impact on Broiler Grow-out Operations." Applied Economic Perspectives and Policy (33): 79-99.

Miller, Thomas A., Gordon E. Rodenwald, and Robert G. McElroy. 1981. Economies of Size in Crop Farming. U.S. Department of Agriculture, Economics and Statistics Service, AER- 472. July.

Mosheim, Roberto, and Lovell, C. A. Knox. 2009. "Scale Economies and Inefficiency of U.S. Dairy Farms." American Journal of Agricultural Economics (91): 777-794.

National Academy of Sciences. 2010. Impact of Genetically Engineered Crops on Farm Sustainability in the United States. National Academies Press. Washington, DC.

Nickerson, Cynthia, Robert Ebel, Allison Borchers, and Fernando Carriazo. 2011. Major Uses of Land in the United States. U.S. Department of Agriculture, Economic Research Service, EIB-89. December.

O'Donoghue, Erik J., Robert A. Hoppe, David E. Banker, Robert Ebel, Keith Fuglie, Penni Korb, Michael Livingston, Cynthia Nickerson, and Carmen Sandretto. 2011. The Changing Organization of U.S. Farming. U.S. Department of Agriculture, Economic Research Service, EIB-88. December.

O'Donoghue, Erik, Michael J. Roberts, and Nigel Key. 2009. "Did the Federal Crop Insurance Reform Act Alter Farm Enterprise Diversification?" Journal of Agricultural Economics (60): 80-104.

Olmstead, Alan L., and Paul W. Rhode. 2001. "Reshaping the Landscape: The Impact and Diffusion of the Tractor in American Agriculture, 1910-1960," Journal of Economic History (61): 663-698.

Rasmussen, Wayne D. 1968. "Advances in American Agriculture: The Mechanical Tomato Harvester as a Case Study." Technology and Culture (9): 531-543.

Rasmussen, Wayne D. 1962. "The Impact of Technological Change on American Agriculture, 1862- 1962." Journal of Economic History (22): pp. 578-591.

Reimund, Donn A., J. Rod Martin, and Charles V. Moore. 1981. Structural Change in Agriculture: The Experience for Broilers, Fed Cattle, and Processing Vegetables. U.S. Department of Agriculture, Economics and Statistics Service, TB-1648. April.

Roberts, Michael J., and Nigel Key. 2008. "Agricultural Payments and Land Concentration: A Semiparametric Spatial Regression Analysis." American Journal of Agricultural Economics (90): 627-643.

Scherer, F.M., and David Ross. 1990. Industrial Market Structure and Economic Performance. Boston: Houghton Mifflin.

Schimmelpfennig, David, and Robert Ebel. 2011. On the Doorstep of the Information Age: Recent Adoption of Precision Agriculture. U.S. Department of Agriculture, Economic Research Service, EIB-80. August.

Schmitz, Andrew, and David Seckler. 1970. "Mechanized Agriculture and Social Welfare: The Case of the Tomato Harvester." American Journal of Agricultural Economics (52): 569-577.

Shields, Dennis. 2012. Federal Crop Insurance: Background. Congressional Research Service. CRS Report to Congress. R40532. December.

Sneeringer, Stacy, and Nigel Key. 2011. "Effects of Size-Based Environmental Regulations: Evidence of Regulatory Avoidance." American Journal of Agricultural Economics (93): 1189-1211.

Street, James H. 1957. "Mechanizing the Cotton Harvest." Agricultural History (31): 12-22.

U.S. Department of Agriculture, National Agricultural Statistics Service (NASS). 2009. Farms, Land in Farms, and Livestock Operations 2008 Summary. February.

Westcott, Paul C., and Michael J. Price. 2001. Analysis of the U.S. Commodity Loan Program with Marketing Loan Provisions. U.S. Department of Agriculture. Economic Research Service. Agricultural Economics Report No. 801.

Whatley, Warren C. 1983. "Labor for the Picking: The New Deal in the South." Journal of Economic History (43): 905-929.

White, T. Kirk, and Robert A. Hoppe. 2012. Changing Farm Structure and the Distribution of Farm Payments and Crop Insurance. Economic Research Service, EIB-91. February.

GLOSSARY

Acre: An area equal to 43,560 square feet or 4,840 square yards.

Cropland: Land that is suitable for or used for crops. In USDA data, cropland includes cropland harvested, cropland used only for pasture or grazing, cropland on which all crops failed or were abandoned, cropland in cultivated summer fallow, and cropland idled or used for cover crops or used for soil improvement but not pastured or grazed.

Family farm: In ERS analyses, any farm for which the principal operator, and persons related to the principal operator by blood or marriage, own more than 50 percent of the farm business.

Farm: In USDA data, any place from which at least $1,000 worth of agricultural commodities were produced or sold, or normally have been sold, in a year.

Farm operator: A person who operates a farm, either doing the work or making day-to-day operating decisions for the farm.

Genetically engineered (GE) crop varieties: GE techniques allow a more precise alteration of a plant's genetic makeup than is possible under conventional plant breeding, and permit targeting single genes in a plant's DNA. Genetic engineering has led to the introduction of traits for herbicide tolerance (HT) in alfalfa, corn, cotton, sugar beets, and soybeans, and for resistance to certain insect pests (Bt) in corn, cotton, and papaya.

Harvested cropland: In USDA data, cropland from which crops were harvested and hay was cut, land used to grow short-rotation woody crops, orchards, citrus groves, Christmas trees, vineyards, nurseries, and greenhouses.

Labor-saving innovation: An innovation that reduces the amount of labor needed for any given amount of output. Innovations may be mechanical, material, biological or organizational. Strictly speaking, it is distinct from input substitution, in which other inputs are substituted for an input whose price rises, in an amount sufficient to maintain output.

Mean: For a dataset, the mean is the sum of the values divided by the number of values. The mean size of a farm, measured by cropland, is the sum of all cropland acres divided by the number of farms with cropland.

Median: The numerical value separating the top half of a distribution from the bottom half—the 50th percentile of values. For cropland, the median farm size is that value which splits the distribution of farms: half of all farms have at least that value of acres, and half have no more.

Midpoint: In this report, the midpoint acreage for cropland is that value for which half of all cropland is on farms that are at least that large, and half of all cropland is on farms that are no larger than that size. Technically, it is the median of the distribution of cropland by farm size.

Precision agriculture: A set of management practices and information technologies (IT), often incorporated in farm equipment, that measure and help to manage intra-field variations in soil attributes, pest presence, product attributes, and production outcomes.

Principal farm operator: The person primarily responsible for day-to-day management decisions on the farm.

End Notes

[1] An earlier ERS report (MacDonald and McBride, 2009) analyzed consolidation in livestock agriculture. Hoppe et al. (2010) document shifts of production to larger farms, using inflation-adjusted sales, while O'Donoghue et al. (2011) document broad changes in U.S. farm structure.

[2] The 2007 Census of Agriculture recorded 406 million acres of cropland and 310 million acres of harvested cropland. Of the non-harvested cropland, 7 million acres represented planted cropland on which the crops failed, but most fell into three other categories: cropland in summer fallow (15 million acres); cropland used only for grazing or pasture (36 million acres); and cropland that was idled or used for cover crops or soil improvement (38 million acres). Land enrolled in the Conservation Reserve Program (CRP), Wetlands Reserve Program (WRP), or other conservation programs may be reported in harvested cropland, idle cropland, or woodland not pastured, depending on its use.

[3] The mean cropland measures are reported in Census of Agriculture publications, but the weighted median requires confidential record-level data, accessed by ERS under an agreement with NASS.

[4] The data can be used in another way to summarize consolidation. In 1982, 184,030 farms were as large as the midpoint farm size, and so operated half of U.S. cropland. In 2007, 86,531 farms were as large as the midpoint. Those "top-half" farms operated 1,215 acres of cropland, on average, in 1982 and 2,350 acres in 2007.

[5] Acreage also shifted among field crops. Cotton and oats together combined for about 80 million acres (about 40 million each) in the late 1920s, compared to 12 million in 2007. The reduction in cotton acreage reflected slowing cotton demand, due to the introduction and

spread of manmade fibers, combined with diminishing yields from traditional production areas in the Southeast and sharply rising yields in newer production areas in the Plains and West (Larson and Meyer, 1996). The reduction in oats reflected the 20th century shift from horses and mules, which were fed oats and hay, to tractors for farm work.

[6] This list expands on that provided in O'Donoghue et al. (2011), who also summarize several other elements of consolidation.

[7] The farm size distribution has become noticeably more skewed in fruits and vegetables, and midpoints provide information that is not apparent in simple means. Tomatoes provide an extreme but telling example. The 2007 census counted 25,809 farms, with 442,225 harvested acres of tomatoes, compared to 14,366 farms and 414,624 acres in 1997. Almost all of the increase in farm numbers occurred among farms with less than an acre of tomatoes: with new procedures designed to capture more very small farms, the 2007 census counted 17,536 farms with less than an acre of tomatoes, compared to 7,407 in 1997. The mean farm size—total acreage divided by the number of farms—fell from 29 to 17 acres, even as the midpoint acreage rose from 589 to 820 acres (table 3).

[8] The rate of return on equity measures the return on capital in the farm business (total assets minus total debt, or farm net worth). The numerator is net farm income, minus an adjustment for the opportunity cost of the unpaid labor and management time provided by farm operators and other unpaid workers. ERS researchers used class aggregates—total returns by all farms in a class divided by total net worth—to reduce the effects of extreme outliers, observations with near zero values of net worth, in financial data.

[9] On version 1 of the annual ARMS, respondents are asked to report hours worked on the farm by the farm's operators, the primary operator's spouse if he or she is not an operator, other unpaid labor (such as children, or spouses of secondary operators), and paid workers. The sample is, therefore, a subsample of the farms in table 5, with 2,087 grain and soybean farms and 1,471 fruit and vegetable operations.

[10] The Census of Agriculture data in figure 3 are based on all States, while the ARMS estimates are based on much smaller samples and exclude Alaska and Hawaii. However, ARMS estimates for the overlap years of 1997, 2002, and 2007 are quite close to Census measures. In 2007, the the midpoint acreage for Census was 1,105 acres, while that from ARMS was 1,093 acres.

[11] The data are drawn from version 1 of the 2010 ARMS (Phase III), where local producers were farms that had direct sales to consumers for human consumption, or that provided crops directly to retail outlets that in turn sold directly to consumers. Interest in organic products is also growing, but organic crop production appears to be concentrated on large farms; the midpoint acreage for farms with certified organic acreage was 1,968 acres in the 2010 ARMS.

[12] More specifically, the concept relates to the period of time long enough to allow the firm to vary all inputs.

[13] The term "pecuniary economies of scale" describes circumstances in which larger firms can obtain inputs at lower prices. In agriculture, such circumstances can apply when buying in large bulk quantities leads to lower unit prices, because the unit costs of shipping, storing, or processing a large order are lower. Examples include purchases of fertilizer and other farm chemicals, and purchases of transportation for products. While such pecuniary economies of scale can provide larger operations with advantages, they do not appear to have become more important, and do not appear to be important drivers of recent changes in farm size.

[14] Because the Agricultural Equipment Manufacturers Association began reporting sales of tractors of 500 horsepower or greater for the first time in 2005, 2005-2011 sales are reported. The fact that the association did not report that category before 2005 itself indicates the shift to much larger and faster tractors.

[15] Cropland density is the weighted average share of land in cropland, averaged across counties, where each county is weighted by its share of the State's cropland. The measure is designed to capture density in areas where the cropland is located.

[16] Pests often thrive on one crop, but not others. Farmers can, therefore, control pest populations by planting a succession of different crops in a field over time (rotation) or by planting a mix of different crops across a field (interplanting).

[17] GE seeds were also widely used on canola, papaya, and sugar beet acreage, and on some squash and sweet corn acreage, in recent years. However, these are minor U.S. crops, and the vast bulk of GE seed acreage is in corn, cotton, and soybeans.

[18] They used ARMS Phase II data, which elicit detailed information on activities carried out with respect to that field, including hours of labor expended in each activity.

[19] ARMS questionnaires for Phase III, version 1, included questions about GE seed use and about labor hours in those years.

[20] Over time, areawide use of Bt seeds may have reduced insect infestations for all corn, cotton, and soybean farms, thereby eliminating any differential labor-saving effect for Bt adopters.

[21] Note that estimates are not reported for all years. USDA has not performed annual comprehensive surveys of tillage practices; instead, the ARMS covers field level practices of specific crops in different years.

[22] See Gardner (2002) for more on increased specialization of U.S. farms during the 20th century.

[23] The table reports estimates of the value of production for the 18 categories, and to construct this table ERS analysts drew on data for harvested acreage elicited in ARMS Phase III questionnaires for 21 crop categories: barley; canola; corn (grain and silage combined); cotton; dry edible beans/peas/lentils; fruits, nuts, and berries; hay (alfalfa and all other combined); nursery and greenhouse crops; oats; peanuts; potatoes; rice; sorghum (grain and silage combined); soybeans; sugarcane; sugarbeets; tobacco; vegetables (processing and all others combined); wheat; other oilseeds; and all other crops.

[24] The increase in mean farm size does not mean that individual tobacco farms all increased acreage; rather, many very small farms shut down, and tobacco acreage and production shifted toward larger farms in different locations.

[25] S. 954, the Agriculture Reform, Food, and Jobs Act of 2013 (being considered in Congress as of this writing), would eliminate DPs and CCPs, while retaining marketing assistance loans and expanding crop insurance.

[26] See Gardner (2002), pp. 254-271 for a discussion.

[27] Half of cropland in the zip code is on farms that are at least as large as the midpoint, and half is on farms that are no larger.

[28] These growth rates are measured as the natural logarithm of the ratio of beginning and end of period value, whereas the growth rates in figure 4 and the associated text are expressed as a percentage of the beginning value. The specific comparisons are for zip codes sorted into six classes, based on payments per acre.

[29] The analysis controlled for initial farm size as well as attributes of the farm's organization. It also used county fixed effects, so that higher levels of payments per acre, relative to the county's average, resulted in higher growth and survival rates, again relative to a mean for the county.

[30] The analysis relied on Census of Agriculture data, which records days worked off the farm, but does not record on-farm labor hours. Hence, the analysis focused on off-farm employment.

[31] The cattle feeder is Cactus Feeders (www.cactusfeeders.com); the dairy operation is Three Mile Canyon Farms (www.threemilecanyonfarms.com); and the sugar business is U.S. Sugar (www.ussugar.com); data on each are drawn from their websites, and not from USDA data. The California firm is the JG Boswell Company, and the information is taken from the 2009 *Western Farm Press* obituary for founder JG Boswell.

[32] A farm operator makes day-to-day management decisions for the farm business. A farm may have multiple operators. ARMS questionnaires ask for a principal operator, and ask whether that operator and persons related by blood or marriage own more than 50 percent of the farm business, where the business consists of assets owned by the farm— including land; equipment and structures; inventories of livestock, production inputs, and harvested crops; and liquid assets such as cash.

[33] See the 2006-2011 annual reports for each company, in sections referring to "business" and "properties."

[34] See MacDonald and McBride (2009), Key and McBride (2007) for hogs, MacDonald and Wang (2011) for broilers, and Mosheim and Lovell (2009) for dairy.

[35] Cropland values are from *Land Values: 2012 Summary* (USDA National Agricultural Statistics Service), while the equipment and structures estimates are averages for like farms drawn from the 2011 ARMS, Phase III.

In: Farms in the United States
Editor: Janine Grosso

ISBN: 978-1-63483-667-8

Chapter 2

WORKING THE LAND WITH 10 ACRES: SMALL ACREAGE FARMING IN THE UNITED STATES*

Doris J. Newton

ABSTRACT

According to the U.S. Census of Agriculture, 294,000 farms operated on 10 or fewer acres in 2007. While most small acreage (SA) operations did very little farming, approximately 50,000 SA farms had gross sales of $10,000 or more in 2007; 3,600 reported grossing $500,000 or more. Thus, a limited land base does not necessarily translate into low sales. This report focuses on SA farms, especially those grossing $10,000 or more in a given year, and examines such characteristics as production strategies, types of products, sales, household income, and financial performance. The 46,000 SA farms with sales between $10,000 and $500,000 in 2007 were primarily active in floriculture, tree nurseries, and fruits and vegetables, while the 3,600 SA farms with at least $500,000 in sales were primarily active in confined livestock production. SA farms produce only small amounts of field crops, such as wheat, corn, or cotton, which require larger amounts of land.

---

* This is an edited, reformatted and augmented version of Economic Information Bulletin Number 123, issued by the U.S. Department of Agriculture, Economic Research Service, April 2014.

Keywords: small acreage farming, farm structure, farm financial performance, farm operator household income, 10 acres or less, women farmers, women principal and secondary operators, small acreage farm characteristics, small farms

What Is the Issue?

Roughly 13 percent of U.S. farms (294,000) operated on 10 acres or less in 2007, and while most of these small acreage (SA) farms generated very little, if any, farm production, some managed to attain sizeable sales despite their limited land base. Given strong and growing empirical evidence that larger farms are more efficient at producing most farm products, what factors account for the apparent continuing financial viability of some small acreage farms? This report looks at small acreage farms having gross sales of $10,000 or more in a given year to better understand the product choices and strategies used by small acreage farms that appear to be operating profitably.

What Did the Study Find?

Almost one-sixth of all U.S. farms in 2007 were SA farms with 10 or fewer acres. These operations controlled approximately 1.7 million acres of farmland (only 0.18 percent of the 922.1 million total acres of U.S. farmland in 2007). However, total sales of all SA farms were approximately $9 billion in 2007, or 3 percent of total U.S. farm sales.

Most SA farms in 2007 produced very little, if any, farm products; yet, 17 percent of SA farms (50,000) had gross sales of at least $10,000. Approximately 122,000 operations, or 42 percent of all SA farms, reported sales under $1,000 in 2007, while 41 percent reported sales between $1,000 and just under $10,000.

Small acreage does not necessarily mean small sales. Over 30,000 SA farms had sales between $10,000 and $50,000 in 2007, while 6,000 SA farms grossed over $250,000 and 3,600 had sales of at least $500,000. These farms tended to specialize in a single stage of the production process when raising livestock, or they produced high-value crops. They produced very little or no field crops, which require large acreages.

Product specializations varied with the size of the farm. The 3,600 farms in the largest SA sales class, at least $500,000 in sales in 2007, consisted primarily of confined hog and poultry operations. But the 46,000 farms with sales between $10,000 and $500,000 in 2007 focused primarily on high-value crops like floriculture, tree nurseries, orchards, and vegetables.

Small acreage (SA) farms and sales share by product specialization group, 2007

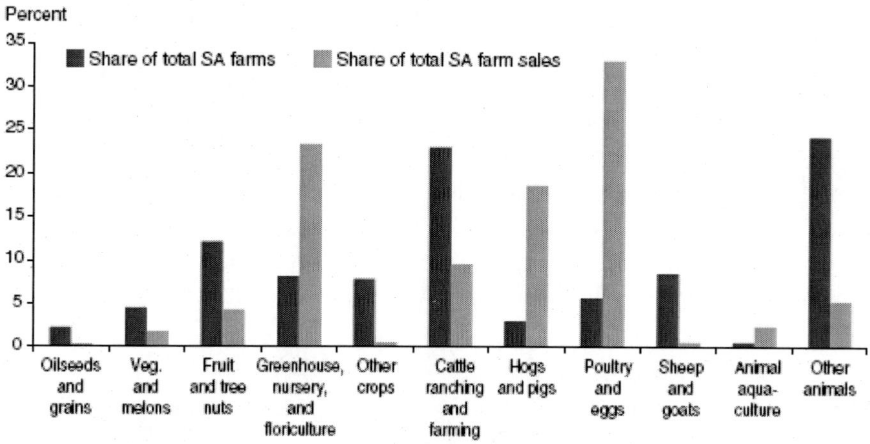

Sources: USDA, Economic Research Service using data from U.S. Census of Agriculture, 2007, and North American Industry Classification System.

Most SA farms with sales greater than $10,000 in 2007 had positive net farm income. The majority of SA farms with sales greater than $10,000 in 2007 realized gross income in excess of cash expenses, and the share with positive returns increased with sales class, except for farms with sales of at least $500,000. Cash expenses include the expense of hired labor but do not account for family-provided labor.

Rates of return on equity for SA farms varied widely among sales classes. ERS estimates of the rate of return on equity account for the costs of family-provided labor. Farms with sales of less than $100,000 in 2007 had negative rates of return, while farms with sales of $100,000 or more had positive rates of return.

The share of women principal operators is higher for SA farms than for all farms. Women were principal operators of 24 percent of all SA farms in 2007, compared with 14 percent of all farms. They were the principal operator on 19 percent of SA farms with sales of $10,000 or more, compared with 8 percent of all farms with sales that high.

How Was the Study Conducted?

This report uses data from both the 2007 U.S. Census of Agriculture and USDA's 2007 Agricultural Resource Management Survey (ARMS) to examine the socioeconomic characteristics of the farming and ranching population whose production is on 10 acres or less. Every 5 years, the agricultural census collects information from farm operators on acres operated, product sales and volumes, operator characteristics, and farm finances. The 2007 census contains the latest available data on all farms at the time this research was completed, and it provides comprehensive details on production by SA farms. ARMS, an annual survey sponsored by USDA's Economic Research Service and National Agricultural Statistics Service, generates a representative national sample of 20,000-22,000 farms and provides additional farm and farm household financial information that is not collected in the census.

Introduction

Is it possible to successfully operate a farm on 10 acres or less? According to the U.S. Census of Agriculture, roughly 294,000 U.S. farms reported operating on 10 acres or less in 2007, the latest year for which data were available for this analysis. For purposes of this report, these farms are referred to as small acreage (SA) farms.

USDA defines a farm as any place from which $1,000 or more of agricultural products (crops and livestock) were produced or sold or normally would have been sold during the year under consideration. USDA's definition does not specify a minimum amount of acreage needed to qualify as a farm, but local government zoning ordinances and State farm support programs may implicitly or explicitly target farms based on the amount of farmland they operate. And for many individuals, a "small farm" may be defined more by the physical size of the operation rather than its financial status. While the amount of crops and livestock that can be produced is a function of the amount of land available to produce it on, the relationship depends on the type of product and choice of farming practices applied. Most SA farms generate very little, if any, farm production in a given year. However, based on the Census of Agriculture, 50,000 SA farms reported gross sales of at least $10,000 in 2007, and 3,600 reported sales of $500,000 or more. This report examines the characteristics of SA farms and the operators and households running them, with a particular focus on SA farms with gross sales of $10,000 or more in a given year.

USDA's Economic Research Service (ERS) has ongoing research that focuses on farm viability. In recent years, ERS has reported on the growing consolidation of production on larger farming operations and the economies of size pushing that trend (MacDonald et al., 2013; O'Donoghue et al., 2011), the aging farm population and trends in numbers of young, beginning farmers (Ahearn, 2013), and the role that small (sales) farms play in the agricultural sector (Hoppe et al., 2010). This report seeks to further this research by examining the role that small acreage farms play in agricultural production, specifically looking at farms that reported having operated on 10 or fewer acres of total farmland devoted to crop and livestock production in 2007. Increasing farmland prices in recent decades (Nickerson et al., 2012) have raised the cost of owning and renting a farm of any appreciable size, meaning some beginning farmers with limited access to land and capital may need to start small. At the same time, farm business viability is a function of production and sales, putting pressure on beginning farmers who want to make farming their primary occupation to find ways of increasing production. This report explores how some SA farm operators have successfully managed their farming business on a limited land base.

Farm Characteristics: General and Small Acreage Farms

Over the last three decades, the size distribution of U.S. farms has changed significantly. Whether farm size is measured by sales or acreage, there has been an increase in the number of smaller and larger farms and a corresponding decline in midsized farms. Production has shifted steadily from midsized to larger farms. Between 1982 and 2007, the number of farms with less than $10,000 in sales (measured in 2007 dollars) in a given year and those with $250,000 or more increased while the number of farms in the middle—those with annual sales between $10,000 and $250,000— declined (table 1).[1]

Like sales, acreage has also shifted over time. Between 1982 and 2007, the number of farms with less than 50 acres and the number with more than 2,000 acres increased, while the numbers of farms in all intermediate classes declined. The numbers of farms with 50 to 179 acres and 500 to 999 acres declined roughly 2 percent or more over the period. Among all farm size classes, the largest decline was in farms that had 180 to 499 acres (table 2).

Several factors contributed to this general shift in farm size distribution. Large farms have better financial performance, on average, than small farms (Hoppe et al., 2010; MacDonald et al., 2013). This provides a strong incentive for farms to become larger and for production to shift to larger farms. Technological advances have increased farm productivity (O'Donoghue et al., 2011; Fuglie et al., 2007). Farms have benefited from productivity-enhancing technologies, such as more efficient farm machinery and advanced scientific disease and pest control, and more efficient use of productive resources, such as allowed by the use of production contracts. Scientific and organizational improvements enable farms to manage more acres and animals effectively.

At the same time, the number of very small farms, those with less than $10,000 in annual sales and/or with fewer than 70 acres of farmland, is increasing (Hoppe and Banker, 2010). A growing number of people who prefer a rural lifestyle have evidently found ways of acting on their preferences, including acquiring ownership of (or otherwise controlling) small amounts of cropland or livestock. However, the trend also reflects two methodological elements of farm surveys, which contribute to the increasing number of small farms as reported in the 2007 Census of Agriculture.

First, the farm definition, set at $1,000 in annual farm sales in 1975, is not adjusted for inflation, so more very small operations qualify as farms over time as commodity prices rise (see box "What Is a Farm?"). The Producer Price Index for Farm Products increased by 86 percent between 1975 and 2007. An operation that had $600 in sales in 1975 would not be defined as a farm in that year, but commodity price inflation would have boosted that same place over $1,000 in sales by 2007, qualifying it as a farm that year.

Second, the National Agricultural Statistics Service (NASS), USDA's primary data collection agency, has made substantial efforts in recent years to improve its coverage of small farms, and success in this endeavor has led to increasing numbers of small farms in its counts (NASS, 2011). NASS prepared for the 2007 Census of Agriculture by partnering with community-based organizations and producer groups to improve survey responses from smaller sales farms, including minority farms.

Table 1. Distribution of U.S. farms and farm sales by sales class

	Number of farms		Market value of sales (billion 2007 $)	
	1982	2007	1982	2007
Total	2,240,976	2, 204,793	189.1	297.2
	Farms and sales distribution by sales class (percent)			
Sales class (2007 $)				
Less than 10,000	42.6	59.8	1.8	0.9
10,000-249,999	50.8	30.7	40.8	14.2
250,000-999,999	5.9	7	30	25.7
1,000,000 or more	0.7	2.5	27.4	59.2
Total	100	100	100	100

Note: Dollar figures have been adjusted for inflation so farms are categorized by sales measured in 2007 dollars. Source: USDA, Economic Research Service using data from U.S. Census of Agriculture, 1982 and 2007.

Table 2. Distribution of U.S. farms by acreage size

	Number of farms		Percent change
	1982	2007	
Farms by size (acres):			
1 to 9	187,665	232,849	2.2
10 to 49	449,252	620,283	8.1
50 to 179	711,652	660,530	-1.8
180 to 499	526,510	368,368	-6.8
500 to 999	203,925	149,713	-2.3
1,000 to 1,999	97,395	92,656	-0.1
2,000 or more	64,577	80,393	0.8
Total farms	2,240,976	2,204,792	

Source: USDA, Economic Research Service using data from U.S. Census of Agriculture, 1982 and 2007.

The Small Acreage Farm Landscape

Farming does not always require much land to generate significant sales. Some farms operate on few acres yet have substantial sales. Almost one-sixth of all U.S. farms in 2007 were small acreage farms with 10 or fewer acres (table 3). As a group, SA farms controlled approximately 1.7 million acres of farmland (0.18 percent of the 922.1 million total acres of U.S. farmland in 2007). However, total sales of all SA farm operations were approximately $9 billion in 2007, or 3 percent of total U.S. farm sales.

What Is a Farm?

USDA defines farms very broadly in its statistical programs. Since 1850, when minimum criteria defining a farm for census purposes were first established, the farm definition has changed nine times as the Nation has grown. A farm is currently defined, for statistical purposes, as any place from which $1,000 or more of agricultural products (crops and livestock) were produced or sold or normally would have been sold during the year under consideration. This definition has been in place since August 1975—by joint agreement between USDA, the Office of Management and Budget, and the Bureau of the Census—and the per year sales amount has not been adjusted for inflation. According to the most recently available (2007) Census of Agriculture, just over 2.2 million farms operate in the United States. Approximately 689,000 farms (31.3 percent) sold less than $1,000 of agricultural products in 2007. These operations were defined as farms, even with sales under $1,000, because they received government farm program payments sufficient to put them over the sales threshold, or because they had cropland or livestock assets sufficient to generate sales of $1,000 had the farms produced and sold commodities from those assets.[1]

[1] If a place does not have $1,000 in sales, a "point system" assigns values for acres of various crops and head of various livestock species to estimate a normal level of sales. Point farms are farms with fewer than $1,000 in annual sales but with points worth at least $1,000. Point farms tend to be very small. Some, however, may normally have large sales but experience low sales in a particular year due to bad weather, disease, or other factors. Both the Agricultural Resource Management Survey (ARMS) and the Census of Agriculture use the point system to help identify farms meeting the current definition.

Who are SA farmers? Not surprising, the majority (77 percent) of SA principal operators in 2007 were men (fig. 1). This finding is consistent with the distribution of farm operators in general, where men were the majority (86 percent). The share of women operators, however, was larger for SA farms than for all farms (fig. 2; see box "Women-Operated Small Acreage Farms").

Table 3. General characteristics of small acreage (SA) farms compared to all U.S. farms, 2007*

	All SA farms	Share of SA totals having $10,000 or more in sales	All U.S. farms	SA share of U.S. totals
		Percent		*Percent*
Number of farms	294,000	17	2,204,792	13
Acreage (millions)	1.7	16	922.1	0.18
Sales ($ billions)	8.9	95	297	3

Source: USDA, Economic Research Service using data from U.S. Census of Agriculture, 2007.

* One-fifth of SA farms had gross annual sales of $10,000 or more and accounted for 95 percent of SA sales in 2007.

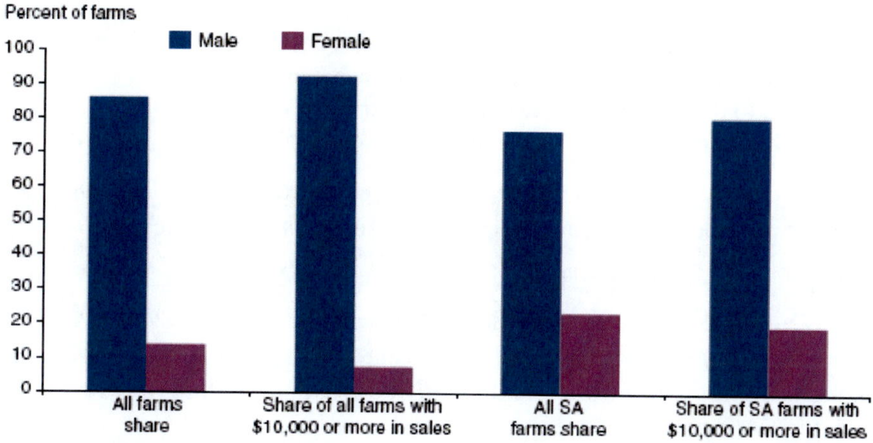

Source: USDA, Economic Research Service using data from U.S. Census of Agriculture, 2007.

Figure 1. Small acreage (SA) and all farms with sales of $10,000 or more in 2007, by gender of principal operator.

Farmers in the United States are aging. The average farmer's age in 2007 was 57 years, and a quarter of all farmers were age 65 or older. SA farmers were slightly younger than the average age. Most SA farmers were age 45 or older, though 22 percent were age 65 or older—similar to the shares for the general farm population (figs. 3 and 4). Among all principal SA farm operators, similar shares of men (25 percent) and women (21 percent) reported being retired.

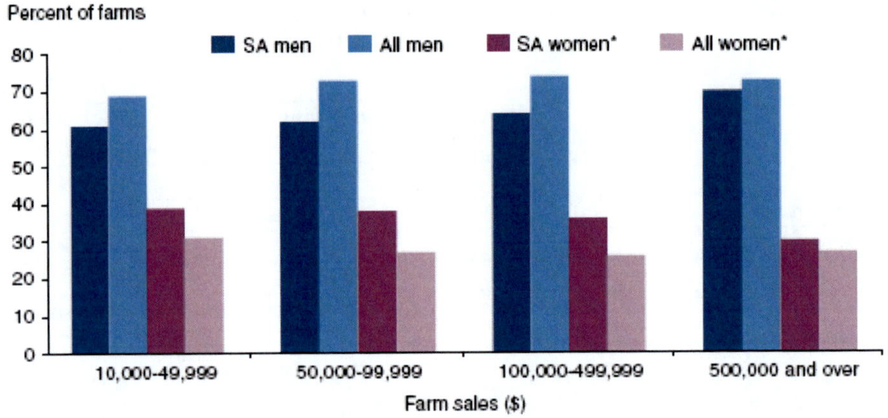

Source: USDA, Economic Research Service using data from U.S. Census of Agriculture, 2007.

* Includes women counted as principal and secondary farm operators, only. Census counts for secondary men and all third operators not included.

Farms with $10,000 or more in sales: SA men count = 39,567; SA women count = 24,329. All men count = 818,591; All women count = 330,157.

Figure 2. Small acreage (SA) and all farms with $10,000 or more in sales in 2007, by gender and with count of women principal and secondary operators combined.

Since access to land is important in farming, farm tenure—how land is held—is a major issue. The majority of SA farm operators in 2007 were full owners (90 percent), while about 4 percent indicated that they were part owners, owning part of the land they operate and renting the rest. The remaining SA farmers were renters.

Women-Operated Small Acreage Farms

Women are much more likely to be principal operators of small acreage (SA) farms than all farms combined (see figure 1 on page 5). In both the Census of Agriculture and USDA's Agricultural Resource Management Survey (sponsored by ERS and the National Agricultural Statistics Service, or NASS), information is collected on the person identified as the principal farm operator. NASS defines a *principal operator* as the person primarily responsible for the onsite, day-to-day operation of the farm or ranch business. This person could be an owner, hired manager, cash tenant, share tenant, and/or a partner.

Data by gender for principal farm operators was first collected in the 1978 Census of Agriculture. And since the 2002 Census of Agriculture, NASS has also collected data on up to three *operators* per farm, recognizing that in some cases farming responsibility is shared even though the second and third parties may not be responsible for day-to-day farm operations. Women farmers play a critical role in many farm operations. Most women farmers work with their husbands to help operate the farm. However, some women are principal farm operators. Unless otherwise indicated, this report focuses on the characteristics of the *principal* operator. Because women are often classified as secondary operators, secondary women farm operators are added to the total count of principal women farm operators as part of the analyses.

Women were principal operators of 24 percent of all SA farms in 2007, compared with 14 percent of all farms. They were the principal operator on 19 percent of SA farms with sales of $10,000 or more, compared with 8 percent of all similar sales farms. When principal and secondary women farm operators are counted, women-operated farms accounted for 58 percent of all SA farms in 2007 and 28 percent of all farms in general. Women-operated farms also accounted for 38 percent of all SA farms with sales of $10,000 or more in 2007 and 29 percent of all similar sales farms.

In every sales class, there were higher shares of women farm operators on SA farms with $10,000 or more sales than on all similar sales farms (see figure 2 on page 5). Most women operators of SA farms were in the lowest referenced sales class ($10,000-49,999). Similar to all SA farms, women-operated SA farms tended to produce specialty crops such as vegetables, fruit, tree nuts, and floriculture and raise beef cattle and other livestock such as horses, goats, and sheep.

Small Acreage Farm Sales

For some products, an operator cannot generate much in sales on a small land base. Indeed, most SA farms generated very little, if any, farm production in 2007. Approximately 122,000 operations, or 42 percent of all SA farms, reported sales under $1,000 in 2007, compared with 31 percent of all farms.

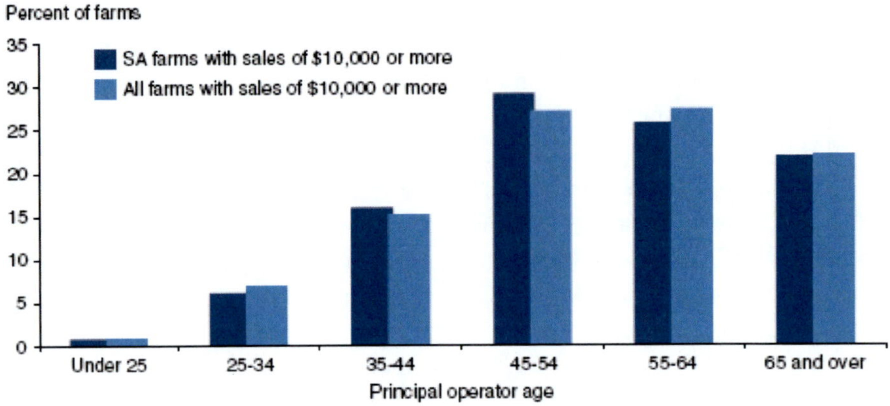

Number of farms = 294,000.

Source: USDA, Economic Research Service using data from U.S. Census of Agriculture, 2007.

Figure 3. All farms and small acreage (SA) farms with $10,000 or more in sales in 2007, by operator age.

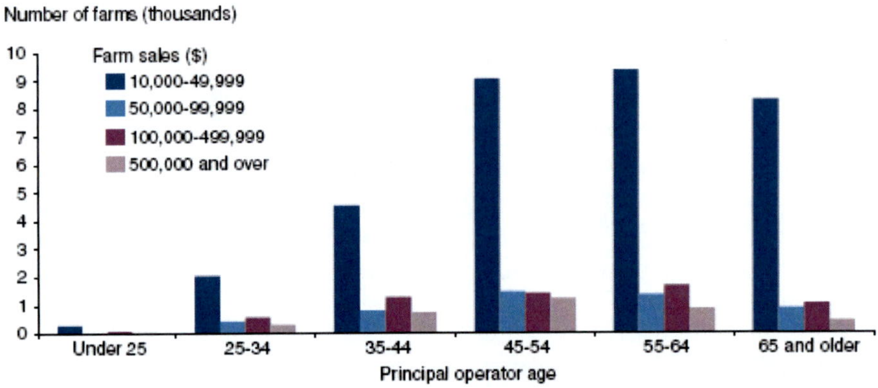

Source: USDA, Economic Research Service using data from U.S. Census of Agriculture, 2007.

Figure 4.Small acreage (SA) farms with $10,000 or more in 2007, by sales category and operator age.

Yet, approximately 50,000 (17 percent) SA farms reported sales of at least $10,000 in 2007 (see table 3), with 3,600 SA farms exceeding $500,000 in sales. To earn over $10,000 in gross sales in a given year, a farm operator has to commit significant resources to farming. For example, in 2007, a livestock farmer would need to have sold 22 feeder cattle at 500 pounds each, at the average 2007 price per animal of $462, to gross over $10,000. The time and effort needed to handle 22 feeder cattle is significant even on a part-time basis. An avocado producer in South Florida would need to sell more than 27,000 pounds of produce, requiring 3-5 acres of bearing trees at recent yields, to get a minimum of $10,000 in gross sales (Evans and Nalampang, 2012).

Acres Operated by Small Acreage Farmers

A relatively high number of SA farms with $10,000 or more in sales in 2007 reported having 1, 5, and 10 acres (table 4), but survey respondents often round numbers in response to questions about acreage. In general, SA farms in this sales class are distributed fairly evenly by the number of acres operated.

Farm Product Specializations and Sales of Small Acreage Farms

If one product or related group of products accounts for at least 50 percent of a farm's total value of production, it is identified as the farm's primary product specialization, according to the North American Industrial Classification System (NAICS) (see appendix table for more details). When a farm does not have one product or one related group of products that makes up 50 percent of the total value of production, it is classified as an "other crops" or "other livestock" operation. The NAICS product group labels are paired with corresponding sales estimates to identify the products of SA farms with annual sales of $10,000 or more, grouped by sales classes. Some SA farms had fairly high sales in 2007. Approximately 3,600 SA farms, or 7 percent of all SA farms, had sales of at least $500,000. SA farms with sales of $500,000 or more in 2007 averaged about $1.6 million in gross sales. SA farms in the intermediate sales classes (from $100,000 to $499,999) averaged roughly $200,000 in annual sales, and those with sales ranging from $50,000 to $99,999 averaged about $68,000. Not surprisingly, SA farms with $10,000 or more in sales accounted for most SA farm sales but just a small share of total SA farms (fig. 5).

The 46,000 SA farms with sales between $10,000 and $500,000 in 2007 were primarily active in floriculture, tree nurseries, and fruits and vegetables, while the 3,600 SA farms with at least $500,000 in sales were primarily active in confined livestock production. More than half of all SA farms specialized in three broad product groups—"other animals" (primarily horses), cattle, and fruit and tree nuts (fig. 6). However, these farm groups combined accounted for only one-fifth of SA sales in 2007. Most SA sales (75 percent) were in three other product groups—poultry and eggs, hogs and pigs, and greenhouse/nursery—though these products were produced on only 15 percent of SA farms. Farm counts and sales were low for SA farms producing grains and oilseeds—field crops that require acreage and input use not typically associated with small farms.

Table 4. Small acreage (SA) farms compared to those with sales of $10,000 or more by acres operated, 2007

Acres operated	SA farms	SA farms with $10,000 or more in sales	Share of SA farms with $10,000 or more in sales
	Number		*Percent*
1	34,510	6,918	20
2	24,768	3,945	16
3	22,319	4,069	18
4	21,232	3,734	18
5	48,902	8,059	16
6	22,639	3,579	16
7	20,957	3,217	15
8	20,653	3,359	16
9	16,869	2,724	16
10	60,826	9,527	16
Totals	293,675	49,131	17

Source: USDA, Economic Research Service using data from U.S. Census of Agriculture, 2007.

Top-Ranking Small Acreage Farm Product Specializations by Sales Class

SA farms with $10,000 or more in sales specialize in a distinct set of farm products that varies with gross sales. For these farms, production does not always require extensive acreage, due in part to the production strategies used. Data on the top 10 products or groups of products produced by SA farms in various sales classes reveal several patterns, specifically the predominance of livestock, floriculture, and other specialty crop farms, particularly as sales increase (table 5; also see appendix table for a list of all SA products for each sales class).

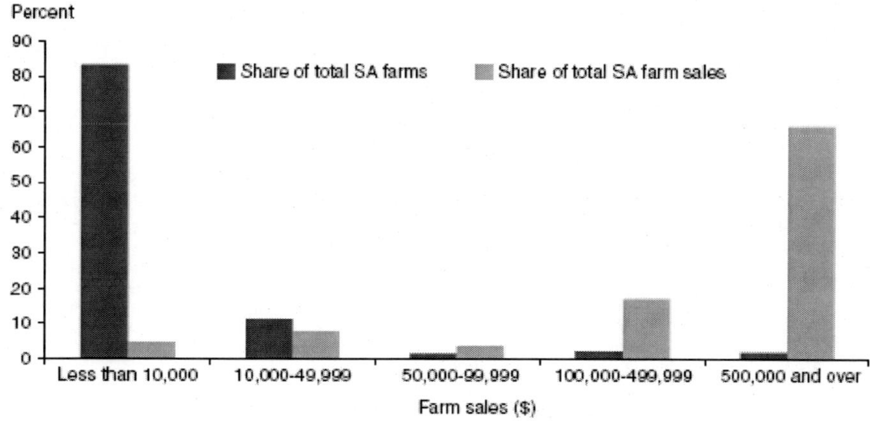

Source: USDA, Economic Research Service using data from U.S. Census of Agriculture, 2007.

Figure 5. Small acreage (SA) farms by sales class in 2007.

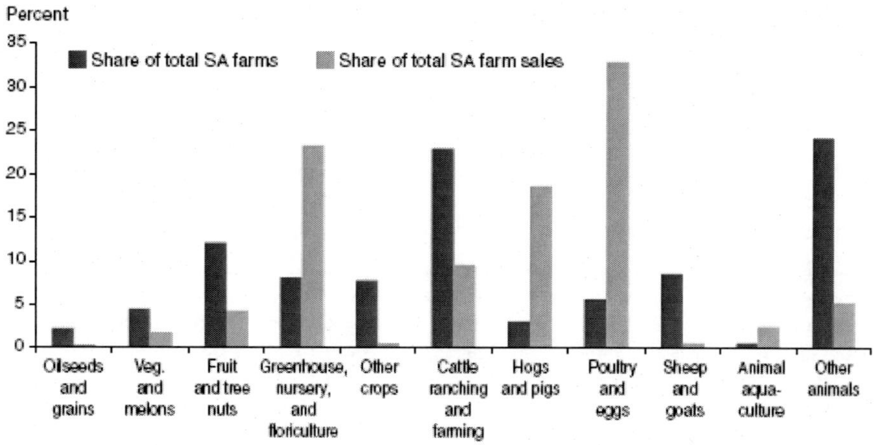

Sources: USDA, Economic Research Service using data from U.S. Census of Agriculture, 2007, and North American Industry Classification System.

Figure 6. Small acreage (SA) farms and sales share by product specialization group, 2007.

Table 5. Top 10 products for small acreage farms with sales of $10,000 or more in 2007, by sales class

Rank	Sales class ($) 10,000-49,999	50,000-99,999	100,000-499,999	500,000 (+)
	Farm count (sales - $ millions)			
1	Floriculture	Floriculture	Floriculture	Hog and pig farming
	3,296 (77)	1,097 (76)	2,000 (421)	909 (1437)
2	Beef cattle	Beef cattle	Nursery/Tree production	Poultry hatcheries
	3,577 (72)	559 (39)	986 (195)	110 (1,066)
3	Other noncitrus fruit	Nursery and tree production	Hog and pig farming	Broilers and other meat type chicken
	3,502 (70)	566 (38)	738 (189)	802 (1,014)
4	Other veg. (except potato) and melon	Other veg. (except potato) and melon	Broilers and other meat type chicken	Floriculture
	3,451 (70)	329 (22)	457 (147)	623 (695)
5	Nursery and tree	Horses/Other equine	Beef cattle	Chicken egg production
	3,048 (65)	307(20)	467 (100)	164(322)
6	Horses/Other equine	Hog and pig farming	Chicken egg	Turkey
	3,223 (64)	265 (19)	337 (96)	155 (237)
	Orange groves	Other noncitrus fruit farming	Dairy cattle/ milk	Nursery and tree
	1,717 (40)	254 (16)	305 (73)	274 (276)
8	Grape vineyards	Apiculture (bee keeping/products)	Finfish farming and fish hatcheries	Mushroom
	1,534 (32)	186 (13)	204 (45)	50 (147)
9	Tree nut farming	Cattle feedlots	Apiculture	Cattle feedlots
	1,342 (26)	134 (10)	239 (45)	55 (128)
10	Apple orchards	Dairy cattle/Milk	Cattle feedlots	Dairy cattle/Milk
	1,087 (23)	138 (10)	152 (35)	105 (147)

Source: USDA, Economic Research Service using data from U.S. Census of Agriculture, 2007, and North American Industry Classification System.
* See appendix table for detailed sales and farm counts by class for all commodities.

Floriculture and mushrooms are high-value specialty products (see box "Selected NAICS 2007 Definitions" for products referenced in this section). Floriculture production ranked number one in all SA sales classes except for the largest ($500,000 or more). Small-acreage floriculture farms can yield high sales because of their high-value products. In addition, floriculture products are often grown under cover (for example, in greenhouses or hoop houses), which allows intensive production over an extended growing season.

Selected NAICS 2007 Definitions

Apiculture - establishments primarily engaged in raising bees. These establishments may collect and gather honey and/or sell queen bees, packages of bees, royal jelly, bees' wax, propolis, venom, and/or other bee products.

Cattle feedlots - establishments primarily engaged in feeding cattle for fattening. The category excludes establishments primarily engaged in operating stockyards for transportation and not buying, selling, or auctioning livestock.

Chicken egg production - establishments primarily engaged in raising chickens for egg production. The eggs produced may be for use as table eggs or hatching eggs.

Floriculture - establishments primarily engaged in growing and/or producing floriculture products (e.g., cut flowers and roses, cut cultivated greens, potted flowering and foliage plants, and flower seeds) under cover and in open fields.

Finfish farming and fish hatcheries - establishments primarily engaged in (1) raising finfish (e.g., catfish, trout, goldfish, tropical fish, and minnows) and/or (2) hatching fish of any kind.

Horses and other equine production - establishments primarily engaged in raising horses, mules, donkeys, and burros.

Oilseed and grain farming - establishments primarily engaged in (1) growing oilseeds, such as flaxseed and cottonseed, and/or grain crops, such as corn and soybeans, and/or (2) producing oilseeds and grain seeds. These crops have an annual life cycle and are typically grown in open fields. This category includes corn silage and grain silage. The 2007 Census of Agriculture classified government-payment only farms as "All Other Crop Farming" (Census of Agriculture, 2007).

Other animal production - establishments primarily engaged in raising animals and insects (except cattle, hogs and pigs, poultry, sheep and goats, and aquaculture) for sale or product production. These establishments are primarily engaged in raising bees, horses and other equines, or rabbits and other fur-bearing animals, or producing goods, such as honey and other bee products. Establishments primarily engaged in raising a combination of animals with no one animal or family of animals accounting for one-half of the establishment's agricultural production (i.e., value of animals for market) are included in this industry group.

Other noncitrus fruit farming - establishments primarily engaged in growing noncitrus fruits (except apples, grapes, berries, and fruit(s) and tree nut(s) combinations). The category includes apricot, avocado, banana, cactus fruit, cherry, coconut tree, coffee, fig, guava, kiwi, mango, nectarine, olive, papaya, passion fruit, peach, pear, persimmon, pineapple, plantain, plum, pomegranate, prickly pear, prune, and quince farming.

Other vegetable (except potato) and melon farming - establishments primarily engaged in one or more of the following: (1) growing melons and/or vegetables (except potatoes; dry peas; dry beans; field, silage, or seed corn; and sugar beets); (2) producing vegetable and/or melon seeds; and (3) growing vegetable and/or melon bedding plants.

Nursery and tree production - establishments primarily engaged in (1) growing nursery products, nursery stock, shrubbery, bulbs, fruit stock, and sod under cover or in open fields and/or (2) growing short-rotation woody trees with a growth and harvest cycle of 10 years or less for pulp or tree stock.

NAICS = North American Industry Classification System. For more information, see www. census.gov/cgi-bin/sssd/naics/naicsrch?chart_code= 11&search=2007

Mushrooms, primarily grown under cover, ranked eighth among products grown by SA farms in the largest sales class in 2007. Mushroom production does not require extensive land; in fact, small-acreage mushroom farms accounted for about 32 percent of all U.S. mushroom farms in 2007 and grossed just over $156 million in sales—about 16 percent of the $985 million in total sales by U.S. mushroom producers (U.S. Census of Agriculture, 2007-Table 37).

Apiculture (beekeeping), aquaculture, and beef production stand out in the midsized $100,000- $499,999 and smaller sales categories. Bees can be kept on small acreages because bee colonies are contained in small, portable structures. Farm-raised catfish and other aquaculture are produced in ponds that do not require large amounts of land. As with SA livestock operations, the feed for fish is generally purchased or provided by the integrator/contractor.

Livestock dominated production among SA farms in the $500,000 or more sales class in 2007. SA livestock operations tend to be highly specialized single-stage producers. The animals are typically in confined feeding areas. Feed is produced offsite and is delivered to the farm by the integrator or parent company. All the animals produced under contract are returned to the parent company. Manure is also usually moved offsite and sold or shared with local farmers or the community.

Beef cattle in feedlots are also in the midsized and lower sales classes. Stockers may raise calves on seasonal pasture, but some farms feed calves from harvested forages in confined or semi-confined settings. In the latter case, some stocker farms use very little land if they confine the animals and rely on purchased forages.

SA farms in the $50,000-$99,999 sales class differ from the larger sales groups because they typically produce vegetables and melons, horses, and noncitrus fruit. The 2007 Census of Agriculture reported over 576,000 horse and pony farms in 2007. Just over 15 percent of these farms oper ated on 10 or fewer acres. About one-quarter of all horse and pony farms sold animals in 2007, compared with over 30 percent of SA farms specializing in horses and ponies. Horses and other equine animals are bought and sold for a variety of purposes, including farm work, breeding, and competitive and recreational riding (AHP, 2009-10). The census data account for riding stables and other equestrian recreational facilities that sell horses and operate like a farm (Offutt and Korb, 2006).

Noticeably absent from the two largest sales classes are SA farms producing "Other Noncitrus" fruit. Fruit in this category are produced seasonally and require seasonal labor and substantial initial capital investments, which could be problematic for many SA farm operators.

In summary, SA farms with sales of $10,000 or more in 2007 produced a broad array of products, including various livestock such as beef cattle, chickens, rabbits, horses, goats, and sheep. However, the majority of sales were primarily from specialized livestock operations with contracts via a parent company. SA farms also produced floriculture, nursery and tree products, and specialty products such as mushrooms, fish, shellfish, and bees and bee byproducts, such as honey. Because of land and equipment constraints, SA farms with sales of $10,000 or more in 2007 did not produce peanuts, rice, dry peas and beans, and oilseeds and other traditional field crops in combination.

Small Acreage Farm Production and Farm Sector Changes

SA farm production is influenced by farm sector changes and the subsequent opportunities available to some small farmers because of the growing importance of production contracts in the livestock industry. Structural changes such as larger farm sizes, improved technologies, increased specialization, and vertical coordination between production stages have enabled some SA farmers to carve out profitable niches with certain products, particularly through the use of production contracts and other organizational arrangements. These changes apply to animal and specialty crop production.

Production contracts are widely used by SA farms that raise hogs, poultry, or replacement dairy heifers, and by fed-cattle operations. Most hog and broiler production is now done in specialized stages. Hog production can be sorted into four distinct stages—breeding and gestation, farrowing, nursery, and finishing. Females are bred and maintained during the first stage. Birth to weaning takes place in the farrowing stage. In the nursery stage, pigs are cared for immediately after weaning until they reach about 30-80 pounds, when they are sent to the finishing stage as feeder pigs. There, the animals are fed to a finishing or slaughter weight of 225-300 pounds (McBride and Key, 2013). Traditional hog farms, known as farrow-to-finish, engage in all four production stages and also raise much of their feed. While most specialized hog finishing operations today raise crops and, hence, have substantial acreage, many finishing operations and most operations that specialize in earlier production stages forego feed crop production for onfarm use and have limited land needs.

Traditional livestock operations use land to raise feed crops for their animals and often use additional land for grazing and pasture. They raise animals from birth until they are ready to be sold for meat. In contrast, specialized operations focus on a single stage of the production process, typically under a production contract. The production contract is an agreement between a farm operation and a firm (or integrator) that sets the terms for transfer of the products from the farm to the integrator or to the integrator's next production stage, such as raising pigs only from farrowing to weaning, feeding yearling cattle to slaughter weights, or raising broiler breeders for eggs to go to hatcheries (the chicks will then be raised for meat on other farms).

In each case, the animals are typically kept in enclosed areas such as open-air pens, if cattle, or in grower houses if broilers and hogs. SA livestock farms with sales of $10,000 or more in a given year often use contracts to link to a larger, more complex operation.

Broiler production complexes include hatcheries, processing plants, and feed mills that are typically owned by firms called integrators. The integrators contract with farmers to "grow out" chicks to market weight for meat, produce eggs for chick production in hatcheries, and raise replacement hens for egg production. The integrator provides contract growers with feed, chicks or hens, and veterinary services. Broiler operations almost always specialize in a single stage. While some are large, diversified crop and livestock operations with extensive acreage, small broiler operators may require only a limited land base for two to three broiler houses and storage facilities for feed and manure if the manure is removed from the operation for use elsewhere as fertilizer. Under production contracts, the contractor typically provides feed and young animals to farmers, who raise the animals and are paid a service fee. Because feed is typically produced offsite, animals are typically confined rather than grazed, and manure is removed from the site, single-stage livestock operations do not require much land (MacDonald and McBride, 2009).

Most dairy farms have crop acreage, but a few purchase all their feed. In that case, it is possible to run a dairy farm with less than 10 acres for cow housing, feed and manure storage, and milking facilities. As in other concentrated feeding operations with no cropland, the manure must eventually be removed for use elsewhere, including as a fertilizer and soil amendment on other farms.

Not all stages of livestock production are conducive to small acreage farming. Beef cattle production includes cow-calf, stocker, and fed-cattle stages. Cow-calf operations consist of a breeding herd, often including a small number of bulls. Most cow-calf operations are land intensive, with cow herds fed from grazing on extensive pasture and hay land. Stocker operations purchase weaned calves from cow-calf operations and raise them until they are moved to feedlots. Stockers may raise calves on seasonal pasture, but some operations feed calves from harvested forages in confined or semiconfined settings. Some stockers may therefore use very little land if they confine the animals and rely on purchased forages, but most use a considerable amount of land for pasture and for feed production.

On fed-cattle operations, feeder cattle are placed in pens with groups of like cattle where they are given grain and special concentrates until they reach a specific size and grade. Most feedlots purchase their feed instead of growing it, so they use land for pens and for feed storage and processing. Small feedlots, with a limited number of pens, need very little land.

SMALL ACREAGE FARM FINANCIAL PERFORMANCE

A farm's financial performance can be gauged by looking at its net farm income and rate of return on equity. This analysis uses data from USDA's 2007 Agricultural Resource Management Survey (ARMS) instead of the census data used elsewhere in this report. ARMS gathers complete financial information needed for measures of farm financial performance, which is unavailable in the census data.

Though small farms in general have negative net farm income, most SA farms with sales of $10,000 or more in 2007 had positive net farm income. Net farm income, however, does not give a complete picture of the profitability of a farm since the expense measure used includes only cash expenses and depreciation and omits the opportunity costs of unpaid farm labor and capital investments. In particular, these omissions can complicate comparisons of financial performance between large and small farms. Large farms are more likely to use hired labor and to incorporate and pay salaries to family members, each of which are recorded as cash expenses. Family labor in farms that are not incorporated does not generate a cash expense; as a result, labor tends to be recorded as an expense in large farms but not in small farms, and failure to recognize this distinction can give a misleading impression of financial performance.

Farmers can often earn money by working off the farm—that is the opportunity cost of the hours that they spend working on their farms. Unless a farm is incorporated and the farmer is paid a salary, the opportunity cost of farm labor is not included as an expense when net farm income is calculated. But more complete measures of farm financial performance are available that account for the costs of using farm operator and other unpaid labor.

The rate of return on equity—a commonly used profitability measure—is net farm income adjusted to account for unpaid labor costs relative to the equity or net worth (assets minus debts) of the farm (see box "Calculating Net Farm Income and Rate of Return on Equity"). It measures the gross rate of return on the farm's invested capital, after accounting for cash expenses and unpaid labor costs.

Average rates of return on equity for SA farms varied widely among sales classes in 2007. Typically, SA farms and all farms with sales of less than $100,000 had negative rates of return—their returns after cash expenses did not cover the opportunity costs of their operators' labor expended on the farm (table 6). Similarly, both types of farms with sales of $100,000 and over in 2007 had positive median rates of return, although SA farms in the $100,000-$499,999 sales class had a higher median rate of return on equity than did all farms in that size class. Given the negative rates of return on equity faced by many farmers, why do they continue to farm? Farmers may continue—despite low rates of return—if they have a positive net farm income and undervalue their labor and management time or if they derive noncurrent income benefits from their farms (eventual capital gains from holding the farm for years). Also, they may receive off-farm income that enables them to continue farming, even if their net farm income is negative.

Calculating Net Farm Income and Rate of Return on Equity

Calculating net farm income:
Net farm income = Gross farm income – expenses

Gross farm income

- Sales of crops and livestock
- *Government payments*
- Other farm-related income (receipts from custom work, machine hire, grazing fees, production contract fees, etc.); change in inventories
- Value of commodities consumed on the farm
- Imputed rental value of the farmhouse

Expenses

- Cash operating expenses
- Depreciation
- In-kind benefits provided to employees

Calculating rate of return on equity: Rate of return on equity = 100× [(net farm income - charge for operator and unpaid labor - charge for management) ÷ net worth]

The rate of return on equity is a commonly used profitability measure that compares net farm income with the net worth of the farm.

In the case of unincorporated farms, net farm income is reduced to reflect the opportunity cost of management and of operator and unpaid labor (see MacDonald et al., 2007, for more information on estimating unpaid labor).

Small Acreage Farm Operator Household Income

Most farm households receive substantial off-farm income from wage-and-salary jobs, self-employment, pensions or Social Security, dividends, interest, and rent. Combined, these off-farm income flows provide the primary source of income for households operating small farms (those with gross annual sales of less than $250,000), as a group.

Table 6. Financial performance of small acreage (SA) and all farms with $10,000 or more in sales in 2007, by sales class

Item	Sales class ($) 10,000 to 49,999	50,000 to 99,999	100,000 to 499,999	500,000 or more	All*
	Number Total farms—family and nonfamily				
SA farms	15,347	3,752	5,492	2,547	226,704
All farms	397,864	148,300	244,670	119,733	2,196,791
	Percent Farms with positive total net farm income				
SA farms	64.2	76.7	81.9	64.7	52.9
All farms	66.7	74.9	81.2	83.5	62.8
	Rate of return on equity (excludes farms with no or negative equity)				
Median-SA	-8.1	-0.2	5.9	3.3	-6.7
Median-All	-2.5	**1.0	1.5	6.7	-2.5

Source: USDA, Economic Research Service using data from USDA's Agricultural Resource Management Survey (ARMS), 2007.

*Includes estimates for farms with sales less than $10,000 that are not shown separately. All farms total is a sample estimate based on ARMS data and differs slightly from estimates for U.S. Census of Agriculture referenced in other tables; based on 871 observations.

**Coefficient of Variation = (Standard Error/Estimate)*100 and is between 25.1 and 50. The coefficient of variation is defined as the standard error of the estimate divided by the estimate, expressed as a percentage. It can be viewed as a measure of statistical reliability of the estimate. Any estimate with a coefficient of variation of 25.1 percent or more is indicated in the tables and figures. Note, however, that estimates close to zero—fairly common among financial ratios—have high coefficients of variation because the denominator of the coefficient of variation is small. In such cases, it is misleading to consider the estimate as unreliable (Ahrendsen and Katchova, 2012).

Only as the farm grows in size (as measured by annual sales) does income from the farm business (farm income, including income from miscellaneous farm-related activities,[2] minus expenses) typically add positively to total farm household income. For all U.S. farm households, median total household income was about $54,000 in 2007 (compared with approximately $50,200 for all U.S. households) (Hoppe et al., 2010). In comparison, the median farm household income for the principal operators of all SA farms was $51,670—less than the median for all farms but above that for all U.S. households. Though median incomes vary when comparing SA farms and all farms with sales of $10,000 or more in 2007, median SA farm household incomes were higher in all sales classes except the largest (table 7).

Table 7. Household income of small acreage (SA) and all farms with sales of $10,000 or more in 2007, by sales class

Item	Sales class ($)				
	10,000 to 49,999	50,000 to 99,999	100,000 to 499,999	500,000 or more	All*
			Number		
Farm households—SA	14,520	3,675	5,417	**1,966	224,322
All	387,081	143,118	236,769	110,152	2,143,398
			Percent		
Distribution of farm households SA	6.5	1.6	2.4	0.9	100
All	18.1	6.7	11.0	5.1	100
			Dollars per household or farm		
Median farm household income					
SA	60,875	75,644	94,218	**145,051	51,670
All	50,277	57,934	78,424	145,051	54,428
Mean farm household income					
SA	84,357	*82,846	**169,260	174,537	72,085
All	78,744	100,034	95,332	262,000	88,796
			Percent		
Median operator farm household income as share of median of all U.S. households (CPS)					
SA	121.2	**150.6	**187.6	288.8	102.9
All	100.1	115.3	156.1	288.8	108.4

Source: USDA, Economic Research Service using data from USDA's Agricultural Resource Management Survey (ARMS), 2007; based on 871 households.

Median income of Current Population Survey (CPS) = $50,233. ARMS 2007 estimates possibly differ from other years because 2007 ARMS weights are calibrated to the 2007 Census of Agriculture. + includes estimates for farms with sales less than $10,000 that are not shown separately. Total households does not include counts of nonfamily farms. **Coefficient of Variation (CV) = (Standard Error/Estimate)*100. * indicates that CV is greater than 25 and less than or equal to 50. The coefficient of variation is defined as the standard error of the estimate divided by the estimate, expressed as a percentage. It can be viewed as a measure of statistical reliability of the estimate. Any estimate with a coefficient of variation of 25.1 percent or more is indicated in the tables and figures. Note, however, that estimates close to zero—fairly common among financial ratios—have high coefficients of variation because the denominator of the coefficient of variation is small. In such cases, it is misleading to consider the estimate as unreliable (Ahrendsen and Katchova, 2012).

Median household income among principal operators of the largest farms was more than twice as high as median household income among SA farms in the $10,000-49,999 sales class. Nonetheless, households operating SA farms in the $10,000-499,999 sales classes had median incomes substantially higher than those of all U.S. farm households with similar levels of farm sales. Households operating SA farms with sales of $10,000 or more are not, in general, low income by national standards.

CONCLUSION AND IMPLICATIONS

Farms can operate as viable businesses on small acreage, but much depends on farm product choices and strategies. Analysis of small acreage farms in this report shows that production strategies, such as contracting, play an important role in livestock production, enabling farmers to operate on limited acreage by specializing in a single production stage. A similar strategy also holds for some specialty crop production, such as floriculture.

SA farms with sales of $10,000 or more in a given year produce a broad array of livestock such as beef cattle, hogs and pigs, dairy, chickens, rabbits, horses, goats, and sheep. They also produce floriculture; nursery and tree products; and specialty products such as mushrooms, fish, shellfish, and bees/bee byproducts such as honey. SA farms of this size are not generally able to produce peanuts, rice, dry peas and beans, and other traditional field crops such as wheat as their primary activity because of the land and equipment needed to generate significant sales of such products.

Four of the top five specializations for SA farms (ranked by sales) are in livestock production, and the fifth is in floriculture. SA livestock operations typically produce under contracts with larger companies (referred to as contractors or integrators), with SA farmers being paid for raising animals provided by the contractor, often with feed and veterinary services provided by the contractor as well.

Despite a nationwide shift in agricultural production to larger farms, small acreage farms with sales of $10,000 or more in a given year are a significant part of the farm landscape. While the majority of small acreage farms do not appear to be operating viable farm businesses, those that do have applied specific production and marketing strategies to achieve at least a modest amount of annual sales, and some produce in excess of $500,000 worth of farm products annually on 10 acres or less.

APPENDIX

Appendix table. Small acreage total farms and sales for all commodity groups by sales class, 2007

	Farm count	Annual sales	Farm count	Annual sales	Farm count	Annual sales	Farm count	Annual sales	All farms	
Commodity group	Less than $50,000		$50,000-$99,999		$100,000-$499,999		$500,000 and more			
All other animal production	10,290	$27,111,384	117	$5,251,240	89	$16,511,606	47	$114,738,895	10,543	$163,613,125
All other grain farming	132	$440,363	None	None	10	$1,977,486	D*	D*	D*	D*
All other miscellaneous crop farming	6,986	$7,490,869	22	$1,461,175	D*	D*	D*	D*	D*	D*
Apiculture	3,714	$21,262,773	186	$13,093,230	239	$44,785,968	15	$10,703,836	4,154	$89,845,807
Apple orchards	3,121	$29,523,711	68	$4,508,989	12	$1,576,368	None	None	D*	D*
Beef cattle ranching and farming	60,085	$165,340,716	559	$38,616,906	467	$99,601,230	108	$112,680,917	61,219	$416,239,769
Berry (except strawberry) farming	2,335	$19,988,393	56	$3,547,648	34	$8,212,962	D*	D*	D*	D*
Broilers and other meat type chicken production	454	$1,357,867	15	$989,650	457	$146,761,087	802	$1,013,684,666	1,728	$1,162,793,270
Cattle feedlots	3,882	$23,236,305	134	$9,998,076	152	$34,948,785	55	$127,954,241	4,223	$196,137,407
Chicken egg production	11,084	$14,311,098	35	$2,592,765	337	$96,267,347	164	$321,892,509	11,620	$435,063,719
Citrus (except orange) groves	1,686	$18,380,636	46	$2,950,558	D*	D*	None	None	D*	D*
Corn farming	3,626	$11,865,217	36	$2,691,448	27	$4,526,490	D*	D*	D*	D*
Cotton farming	34	$131,017	None	None	None	None	None	None	D*	D*
Dairy cattle and milk production	1,600	$18,674,975	138	$9,691,346	305	$72,903,085	105	$146,651,983	2,148	$247,921,389
Dry pea and bean farming	12	$30,349	None	None	None	None	None	None	D*	D*

Appendix table. (Continued)

Commodity group	Farm count	Annual sales	Farm count	Annual sales	Farm count	Annual sales	Farm count	Annual sales	All farms	
	Less than $50,000		$50,000-$99,999		$100,000-$499,999		$500,000 and more			
Finfish farming and fish hatcheries	550	$6,014,206	90	$6,237,210	204	$45,396,528	58	$81,449,202	902	$139,097,146
Floriculture production	7,396	$93,549,258	1,097	$76,077,013	2,000	$420,905,875	623	$694,632,740	11,116	$1,285,164,886
Fruit and tree nut combination farming	312	$1,276,620	None	None	None	None	None	None	D*	D*
Fur-bearing animal and rabbit production	984	$3,337,743	23	$1,588,223	48	$11,225,753	16	$13,279,233	1,071	$29,430,952
Goat farming	13,595	$19,967,305	36	$2,579,832	25	$5,179,120	D*	D*	D*	D*
Grape vineyards	4,491	$42,087,097	131	$8,794,918	30	$4,023,894	D*	D*	D*	D*
Hay farming	14,842	$24,350,112	D*	D*	D*	D*	D*	D*	D*	D*
Hog and pig farming	7,049	$28,771,768	265	$19,478,347	738	$188,965,413	909	$1,436,670,448	8,961	$1,673,885,976
Horses and other equine production	55,057	$118,726,533	307	$19,891,396	146	$25,218,434	25	$28,798,102	55,535	$192,634,465
Mushroom production	57	$705,831	11	$857,413	32	$7,615,725	50	$147,233,420	150	$156,412,389
Nursery and tree production	9,920	$85,592,108	566	$38,370,443	986	$194,827,960	274	$275,726,141	11,746	$594,516,652
Oilseed and grain combination farming	27	$73,362	None	None	None	None	None	None	D*	D*
Orange groves	3,477	$46,745,758	125	$8,033,347	D*	D*	None	None	D*	D*
Other animal aquaculture	66	$1,066,146	11	$680,119	26	$4,540,303	D*	D*	D*	D*
Other food crops grown under cover	631	$12,315,652	99	$6,843,946	78	$16,789,530	26	$24,236,813	834	$60,185,941
Other noncitrus fruit farming	11,812	$99,419,293	254	$16,222,594	48	$6,635,390	None	None	D*	D*
Other poultry production	2,312	$7,653,953	37	$2,612,805	58	$15,054,425	15	$11,937,764	2,422	$37,258,947
Other vegetable (except potato) and melon farming	12,599	$104,056,165	329	$21,548,609	93	$16,399,466	16	$16,536,147	13,037	$158,540,387

Commodity group	Farm count	Annual sales	Farm count	Annual sales	Farm count	Annual sales	Farm count	Annual sales	All farms	
	Less than $50,000		$50,000-$99,999		$100,000-$499,999		$500,000 and more			
Peanut farming	25	$57,712	None	None	None	None	None	None	D*	D*
Potato farming	190	$1,344,892	None	None	None	None	None	None	D*	D*
Poultry hatcheries	203	$512,498	D*	D*	12	$2,729,288	110	$1,066,192,925	D*	D*
Rice farming	11	$45,713							11	$45,713
Sheep farming	11,436	$17,971,381	17	$1,184,344	12	$3,111,934	D*	D*	D*	D*
Shellfish farming	531	$7,304,615	82	$5,643,643	90	$17,796,364	25	$36,553,126	728	$67,297,748
Soybean farming	2,465	$6,906,540	D*	D*	D*	D*	None	None	D*	D*
Strawberry farming	390	$4,748,811	42	$2,903,985	72	$9,689,405	None	None	D*	D*
Sugar beet farming	20	$104,295	None	None	None	None	None	None	D*	D*
Sugarcane farming	10	$50,323	None	None	None	None	None	None	D*	D*
Tobacco farming	986	$10,530,044	10	$582,258	D*	D*	None	None	D*	D*
Tree nut farming	7,009	$40,814,429	17	$1,134,300	None	None	None	None	D*	D*
Turkey production	350	$1,197,476	D*	D*	47	$15,941,795	155	$236,514,891	D*	D*
Wheat production	286	$546,242	None	None	None	None	None	None	D*	D*

Source: USDA, Economic Research Service using data from U.S. Census of Agriculture, 2007.
None=no data existed for this category. D*=disclosure restrictions apply.

References

Ahearn, Mary. *Beginning Farmers and Ranchers at a Glance,* Economic Brief No. 22, U.S. Department of Agriculture, Economic Research Service, January 2013, www.ers.usda.gov/publications/eb-economic-brief/eb22.aspx

Ahrendsen, Bruce L., and Ani L. Katchova. "Financial Ratio Analysis Using ARMS Data," *Agricultural Finance Review*, Vol. 72, Issue 2, 2012, pp. 262-272, www.emeraldinsight.com/journals.htm?articleid=17044258&show=abstract

American Horse Publications (AHP). *Equine Industry Survey, Summary Statistics, 2009-2010*, 2010, www.americanhorsepubs. org/resources/AHP-Equine-Survey-Final.pdf

Cash, A. James, II. "Where's the Beef? Small Farms Produce Majority of Cattle," *Agricultural Outlook*, Issue 297, U.S. Department of Agriculture, Economic Research Service, December 2002, pp. 21-24, http://webarchives.cdlib.org/sw1610ws5g/http://www.ers.usda.gov/publications/agoutlook/Dec2002/ao297g.pdf

Evans, Edward A., and Sikavas Nalampang. "Sample Avocado Production Costs and Profitability Analysis for Florida," University of Florida Extension, 2012, www.edis.ifas.ufl.edu/fe837#

Fuglie, Keith O., James M. MacDonald, and Eldon Ball. *Productivity Growth in U.S. Agriculture*, Economic Brief No. 9, U.S. Department of Agriculture, Economic Research Service, September 2007, www.ers.usda.gov/media/201254/eb9_1_.pdf

Hoppe, Robert A., and David E. Banker. *Structure and Finances of U.S. Farms: Family Farm Report, 2010 Edition*, Economic Information Bulletin No. 66, U.S. Department of Agriculture, Economic Research Service, July 2010, www.ers.usda.gov/publications/eib-economic-information-bulletin/eib66.aspx

Hoppe, Robert A., and Penni Korb. *Characteristics of Women Farm Operators and Their Farms*, Economic Information Bulletin No. 111, U.S. Department of Agriculture, Economic Research Service, April 2013, www.ers.usda.gov/publications/eib-economic-information-bulletin/eib111. aspx

Hoppe, Robert A., James M. MacDonald, and Penni Korb. *Small Farms in the United States: Persistence Under Pressure,* Economic Information Bulletin No. 63, U.S. Department of Agriculture, Economic Research Service,

February 2010, www.ers.usda.gov/publications/eibeconomic-information-bulletin/eib63.aspx

Jerardo, Alberto. *2007 Floriculture and Nursery Crops Yearbook,* U.S. Department of Agriculture, Economic Research Service, Outlook Report No. FLO-2007, September 2007, http://webarchives. cdlib.org/sw1s17tt5t/http://ers.usda.gov/Publications/flo/2007/09Sep/FLO2007.pdf

MacDonald, James M. *The Economic Organization of U.S. Broiler Production*, Economic Information Bulletin No. 38, U.S. Department of Agriculture, Economic Research Service, June 2008, www.ers.usda.gov/publications/eib-economic-information-bulletin/eib38.aspx

MacDonald, James, and Penni Korb. *Agricultural Contracting Update: Contracts in 2008*, Economic Information Bulletin No. 72, U.S. Department of Agriculture, Economic Research Service, February 2011, www.ers.usda.gov/publications/eib-economic-information-bulletin/eib72. aspx

MacDonald, James M., Erik J. O'Donoghue, William D. McBride, Richard F. Nehring, Carmen L. Sandretto, and Roberto Mosheim. *Profits, Costs, and the Changing Structure of Dairy Farming*, Economic Research Report No. 47, U.S. Department of Agriculture, Economic Research Service, September 2007, www.ers.usda.gov/media/188030/err47_1_.pdf

MacDonald, James M., Penni Korb, and Robert A. Hoppe. *Farm Size and the Organization of U.S. Crop Farming*, Economic Research Report No. 152, U.S. Department of Agriculture, Economic Research Service, August 2013, www.ers.usda.gov/publications/err-economic-research-report/ err 152.aspx

McBride, William D., and Nigel Key. *U.S. Hog Production From 1992 to 2009: Technology, Restructuring, and Productivity Growth,* Economic Research Report No. 158, U.S. Department of Agriculture, Economic Research Service, October 2013, www.ers.usda.gov/publications/erreconomic-research-report/err158.aspx

Nickerson, Cynthia, Mitch Morehart, Todd Kuethe, Jayson Beckman, Jennifer Ifft, and Ryan Williams. *Trends in U.S. Farmland Values and Ownership*, Economic Information Bulletin No. 92. U.S. Department of Agriculture, Economic Research Service, February 2012, www.ers.usda.gov/publications/eib-economic-information-bulletin/eib92.aspx

O'Donoghue, Eric J., Robert A. Hoppe, David E. Banker, and Penni Korb. *Exploring Alternative Farm Definitions: Implications for Agricultural Statistics and Program Eligibility*, Economic Information Bulletin No. 49, U.S. Department of Agriculture, Economic Research Service, March

2009, www.ers.usda.gov/publications/eib-economic-information-bulletin/eib49.aspx

O'Donoghue, Eric J., Robert A. Hoppe, David E. Banker, Robert Ebel, Keith O. Fuglie, Penni Korb, Michael Livingston, Cynthia Nickerson, and Carmen L. Sandretto. *The Changing Organization of U.S. Farming*, Economic Information Bulletin No. 88, U.S. Department of Agriculture, Economic Research Service, December 2011, www.ers.usda.gov/publications/eib-economicinformation-bulletin/eib88.aspx

Offutt, Susan, and Penni Korb. "More Women Turning to Horse Farming," *Amber Waves*, Vol. 4, Issue 4, U.S. Department of Agriculture, Economic Research Service, September 2006, http://webarchives.cdlib.org/sw1vh5dg3r/ http:// www.ers.usda.gov/AmberWaves/September06/ DataFeature/)

Regents of the University of Minnesota. "Commercial Fruit Production in Minnesota," 2007, http:// fruit.cfans.umn.edu/apples/beforeyoustart.htm

U.S. Department of Agriculture, Agriculture Marketing Service. *National Fruit and Vegetable Retail Report, Summary*, Vol. VI, No. 29, pg. 3, July 20, 2012, www.ams.usda.gov/mnreports/ fvwretail.pdf

U.S. Department of Agriculture, Economic Research Service. "Agricultural Resource Management Survey (ARMS)," 2007, www.ers.usda.gov/data-products/arms-farm-financial-and-crop-production-practices.aspx

U.S. Department of Agriculture, Economic Research Service. "Farm Household Well-Being," updated: August 27, 2012, www.ers.usda.gov/ topics/farm-economy/farm-household-well-being/glossary.aspx%23 income#income

U.S. Department of Agriculture, Economic Research Service. "Farm Household Economic Well-Being: Glossary," 2013, www.ers.usda.gov/ topics/farm-economy/farm-household-well-being/ glossary.aspx

U.S. Department of Agriculture, Economic Research Service. *Livestock, Dairy, and Poultry Outlook*, Outlook Report No. LDP-M-154, April 18, 2007, www.ers.usda.gov/publications/ ldp/2007/04Apr/LDPM154.pdf

U.S. Department of Agriculture, Economic Research Service. "Vegetables and Pulse Data-Mushroom Outlook Tables (M-5)," 2007, www.ers.usda.gov/ data-products/vegetables-and-pulsesdata/outlook-tables.aspx

U.S. Department of Agriculture, National Agricultural Statistics Service (NASS). *Farms, Land in Farms, and Livestock Operations, 2010 Summary*, February 2011, pp. 26-27, www.nass.usda.gov/ Publications/ Todays_Reports/reports/fnlo0211.pdf

U.S. Department of Agriculture, National Agricultural Statistics Service (NASS). "2007 Census of Agriculture, Volume 1, Chapter 1: U.S.

National Level Data, 2009, Table 49, Selected Operator Characteristics for Principal, Second, and Third Operator: 2007."

U.S. Department of Agriculture, National Agricultural Statistics Service (NASS). "Small Farms Fact Sheet," www.agcensus.usda.gov/Publications/2007/Online_Highlights/Fact_Sheets/Farm_ Numbers/small_farm.pdf

U.S. Department of Agriculture, National Agricultural Statistics Service (NASS). "Mushrooms, Greenhouse, Nursery, Floriculture Fact Sheet," www.agcensus.usda.gov/Publications/2007/ Online_Highlights/Fact_Sheets/Production/nursery.pdf

U.S. Department of Agriculture, National Agricultural Statistics Service (NASS). "2007 Census of Agriculture," www.agcensus.usda.gov/Publications/2007/Full_Report/Volume_1,_Chapter_1_ US/usintro.pdf

U.S. Department of Agriculture, National Agricultural Statistics Service (NASS). "United States Summary and State Data, Volume 1, Geographic Area Series, Part 51," various, www.agcensus. usda.gov/Publications/2007/Full_Report/usv1.pdf

U.S. Department of Agriculture, National Agricultural Statistics Service (NASS). "United States Summary and State Data, Volume 1, Table 37. Nursery, Greenhouse, Floriculture, Sod, Mushrooms, Vegetable Seeds, and Propagative Materials Grown for Sale: 2007 and 2002," www. agcensus. usda.gov/Publications/2007/Full_Report/Volume_1,_Chapter_1_US/st99_1_037_037. pdf

U.S. Department of Commerce, U.S. Census Bureau. North American Industrial Classification System (NAICS), 2007, www.census.gov/cgi-bin/sssd/naics/naicsrch?chart=2007

End Notes

[1] The adjustment for inflation is done because farm prices received have risen over time, so a farm with the same quantity of production would have higher sales in 2007 than in 1982 when measured in nominal (current) dollars. The inflation adjustment aims to correct for price changes so that we measure changes in real sales growth over time. According to the Producer Price Index for Farm Products, farm prices rose by 43 percent between 1982 and 2007.

[2] Other farm-related activities can include operating another farm, renting out farmland to another operator, and the sale of forest products.

INDEX

D

E

F

G

H

I

L

M

N

O

P

Q

R

S

Y